The philosophy of quantum mechanics

T0282328

The philosophy of quantum mechanics

An interactive interpretation

Richard Healey

The right of the
University of Cambridge
to print and sell
all manner of books
was granted by
Henry VIII in 1534.
The University has printed
and published continuously
since 1584.

Cambridge University Press

Cambridge
New York   Port Chester   Melbourne   Sydney

Published by the Press Syndicate of the University of Cambridge
The Pitt Building, Trumpington Street, Cambridge CB2 1RP
40 West 20th Street, New York, NY, 10011-4211, USA
10 Stamford Road, Oakleigh, Victoria 3166, Australia

First published 1989
Reprinted 1990
First paperback edition 1990
Reprinted 1991

Library of Congress Cataloging-in-Publication Data
Healey, Richard.
The philosophy of quantum mechanics: an interactive
interpretation / Richard Healey
p.   cm.
Bibliography: p.
ISBN 0-521-37105-8
1. Quantum theory.   2. Physics – Philosophy.   1. Title.
QC174.12.H43   1990

530.1'2 – dc19

89-30943
CIP

British Library Cataloguing in Publication Data
Healey, Richard
The philosophy of quantum mechanics.
1. Quantum theory.
I. Title
530.1'2

ISBN 0-521-37105-8 hardback
ISBN 0-521-40874-1 paperback

Transferred to digital printing 2003

For Jean

Contents

Preface   *page* xi

Introduction   1
What I take to be an interpretation of quantum mechanics, and why
any further interpretation is needed: the chief problems with the
most popular existing interpretations (both Copenhagen and non-
Copenhagen).

1 Overview   26
Basic ideas of the interpretation to be presented here, together with
idealized examples illustrating their application, including repeated
Stern-Gerlach "measurements" and coupled spin-½ systems.

2 Dynamical states   63
Dynamical states as collections of quantum dynamical properties.
The structure and evolution of dynamical states.

§2.1 The hierarchy of quantum systems   63
How compound quantum systems are composed of their subsystems, and how their dynamical properties are related to those of
their components. Prime and composite properties.

§2.2 The structure of dynamical states   66
Conditions on properties contained in the dynamical state of a
quantum system. How these conditions relate to the hierarchy of
quantum systems, and to the presence or absence of interactions.

§2.3 Dynamics   79
The dynamical law underlying the Schrödinger equation, and a
stability condition governing dynamical states of interacting
systems.

vii

3   Measurement and quantum states   84
The nature of quantum states: measurement, preparation, and the derivation of the Born probability rules. Dissolution of the measurement problem.

§3.1   Measurement interactions   84
Measurements as quantum interactions. Definition of simple *M*-type interactions. Mimicking of the projection postulate through verifiability of measurement results. Motivation for the distinction between prime and compound properties.

§3.2   Idealizations relaxed   94
How the above account of measurement interactions may be rendered more realistic.

§3.3   The assignment of quantum states   104
The character and legitimacy of quantum state ascriptions. Preparations as quantum interactions. Definition of the class of simple *P*-type interactions. Derivation of the Born rules in special cases.

4   Coupled systems   116
The present interpretation is applied to coupled systems of the type studied by Einstein-Podolsky-Rosen, Bell, and Aspect I: technical details.

5   Metaphysical aspects   137
The present interpretation is applied to coupled systems of the type studied by Einstein-Podolsky-Rosen, Bell, and Aspect II: metaphysical aspects and implications concerning holism, nonseparability, causal explanation, and causation. How quantum mechanics explains the observed properties of these correlated systems according to the present interpretation.

6   Alternatives compared   180
The present interpretation is compared to several existing interpretations in order to clarify its logical and genetic relations to them as well as to point out its advantages.

§6.1   Naive realism   180
Comparison with an interpretation often attributed to Einstein.

§6.2   Copenhagen   184
Comparison with two prominent versions of the Copenhagen interpretation.

§6.3   *Everett*   205

Comparison with Everett's relative state interpretation, popularly known as the "many worlds" interpretation.

§6.4   *Kochen*   216

Comparison with interpretative views put forward recently by Kochen.

7   *Open questions*   235

A number of open questions for the present interpretation are posed and discussed. These include the possibility of generalizing the present treatment of measurement and preparation interactions, of extending this interpretation to a relativistic theory, of investigating the nature of the stochastic dynamics underlying quantum probabilities in the present interpretation, and of proving certain important limit results.

Appendix   253

Containing proofs of several results and lemmas stated and used in the text.

Selected bibliography   261
Index   265

Preface

In this book I present a new interpretation of the theory of nonrelativistic quantum mechanics. In this interpretation, measurement in quantum mechanics corresponds to a physical interaction internal to a compound quantum system, which correlates the dynamical states of measured system and (quantum mechanical) apparatus system; whereas the quantum state provides (via the Born rules) a summary of numerical information concerning probabilistic dispositions realized in such interactions. The distinction between dynamical and quantum states is maintained to be the key to the dissolution of the measurement problem. The further idea that the dynamical state of a compound quantum system is not always determined by those of its components then permits a novel understanding of EPR-type correlations as manifested in experiments such as those of Aspect, Grangier, and Roger (1982a) and Aspect, Dalibard, and Roger (1982b).

This book is a monograph: Our present understanding of quantum mechanics does not suffice for the composition of a treatise. It arose when the continual frustration of my attempts to understand quantum mechanics in accordance with any existing interpretation forced me to try something new. I have no illusions that this monograph finally renders quantum mechanics transparent and manifestly free of all conceptual problems. But I do believe that new ideas are urgently needed if we are to approach this happy state; and that there is a good chance that the particular combination of ideas contained in the present interpretation represents significant progress toward this goal. Consequently, I have written mostly in a spirit of advocacy for the interpretation presented here, developing the view so that its advantages are emphasized. But I have also tried to clearly acknowledge difficulties, particularly in the concluding chapter. I hope that some reader will be motivated by the promise of the view to develop or modify the interpretation so as to overcome such difficulties. After an introductory chapter which explains and motivates the attempt to provide a new interpretation of quantum mechanics, Chapter 1 outlines the central ideas of the new interpretation. The detailed presentation begins in Chapter 2, where the character of dynamical states is investigated in some depth. Chapter 3 analyzes quantum measurement and preparation interactions, explains the role of the quantum state, and discusses the interpretation and origin of the Born probability rules of quantum mechanics. Chapter 4 gives technical details of the application of the present interpretation of quantum mechanics to coupled systems of the type studied by EPR, Bell, and Aspect, and Chapter 5 explores metaphysical aspects and implications of this application concerning holism and causal explanation. In Chapter 6 I explicitly compare the interpretation to certain other interpretations which were influential in its development: The present interpretation represents an evolution from earlier views, not a radical break with them. I conclude in Chapter 7 by discussing a number of problems still to be faced by the interpretation presented here. A few proofs whose presentation in the text would have interrupted the exposition have been relegated to an appendix.

I first began the work which resulted in this monograph in the winter of 1984 while I was a visiting fellow of the Center for Philosophy of Science at the University of Pittsburgh. I wish to thank the National Endowment for the Humanities, whose award made this visit financially possible, and the Center and its Director Nicholas Rescher and secretary Linda Butera, whose generous hospitality, both then and later (during the academic year 1987–8), provided me with ideal working conditions in a friendly and stimulating intellectual community. After my return to Los Angeles, continued work on the project during the period 1985 through 1987 was facilitated by the NSF. Consequently, this monograph is, in part, based upon work supported by the National Science Foundation under Grant no. SES8512026: It does not, of course, reflect the views of that organization. I am grateful to the NSF for their support. I have had the benefit of presenting various earlier versions of parts of this work before a number of audiences, including the philosophy of physics mini-conference at Stanford University organized by Nancy Cartwright in February 1985; the philosophy departments of Cornell University, the University of California, Santa Barbara, Arizona State University, Rice University, Columbia University, Amherst College, the University of Western Ontario, Pittsburgh University, and the University of California, Davis; and the Joint US–USSR colloquium on the foundations of quantum mechanics at Easton, Maryland, organized by Jeffrey Bub in September 1988. I wish to single out the Relativity Group at the University of Chicago for special mention. Under the active leadership of Robert Geroch, participants at a talk I gave there in April 1984 not only reacted in a constructive and open-minded way to thoughts which at that stage were considerably less than half-baked, but also took pleasure in resolving technical questions in a matter of minutes, a positive answer to which at that time was crucial to the continued development of my interpretation.

I have a number of individuals to thank for constructive criticisms of my work during this period, including David Albert, Geoffrey Hellman, David Malament, Hilary Putnam, Abner Shimony, Allen Stairs, and Roberto Torretti. I am grateful to Howard Stein and Al Janis, both of whom read a penultimate draft of the first five chapters of the manuscript with exceptional speed, care, and thoughtfulness, and offered advice which I have usually taken (and ignored at my peril!).

Earlier in the project, conversations with both Al and Tony Martin were crucial in determining the direction of my thinking; although I am confident that neither of them would endorse the present interpretation, and certainly neither of them can be held responsible for the consequences of their interventions.

My greatest debt is to my family, and especially to my wife, Jean Hampton. As a colleague she has provided constant encouragement and wise philosophical counsel while I have been working on this project. As a friend she has been a source of emotional support during some difficult times. But the countless sacrifices she has made on my behalf, of career opportunities as well as of her own precious time, far exceed those appropriate to any colleague or friend: They constitute a debt which I can never hope to repay, but can only wonder at.

Davis, California
January 1989

Introduction

In this monograph I present the basic structure of an interpretation of quantum mechanics. Chapter 1 outlines the central ideas behind the interpretation and illustrates them by means of examples: Subsequent chapters fill in the details of the interpretation. But it is important to begin by explaining what I take an interpretation of quantum mechanics to be, and why any further interpretation needs to be offered. After all, quantum mechanics (in some form) is by now both a foundation for much of contemporary physics and a veteran of more than sixty years of intermittent but sometimes intense reflection on its content and meaning. What more can be, or should be, said about the interpretation of this theory?

Many physicists believe that no more needs to be said: that there is basically only one way of understanding quantum mechanics, due to Bohr, Heisenberg, Pauli, and others, and that the only remaining interpretative task is that of the physics teacher, who seeks to perfect ways of conveying this understanding to new generations of students. Sir Rudolf Peierls, one of the more lucid and distinguished of these physicists, even objects to the use of the familiar term 'Copenhagen interpretation' to refer to the way of understanding quantum mechanics due to Bohr, Heisenberg, Pauli, and others.

Because this sounds as if there were several interpretations of quantum mechanics. There is only one. There is only one way in which you can understand quantum mechanics. There are a number of people who are unhappy about this, and are trying to find something else. But nobody has found anything else which is consistent

1

yet, so when you refer to the Copenhagen interpretation of the mechanics what you really mean is quantum mechanics. And therefore the majority of physicists don't use the term; it's mostly used by philosophers.[1]

I am a philosopher, and I shall sometimes find it convenient to talk about the Copenhagen interpretation of quantum mechanics. But that may well be the only respect in which I can assent to the views expressed by Peierls in this passage. It seems to me that far from there being only one interpretation of quantum mechanics, there is today *no* fully satisfactory way of understanding this theory. Instead we are faced with an extraordinary variety of attempts to understand quantum mechanics: Indeed, it sometimes seems as if there are as many different attempts as there are people who have seriously made the attempt! But none of these attempts has either won, or deserved, universal or even widespread acceptance. It is sometimes useful to classify these different interpretation-sketches, since they do fall into certain groups. Thus one may refer to the many-worlds interpretation, to an interpretation in terms of hidden variables, to a naive realist interpretation, to the quantum logical interpretation, or to the Copenhagen interpretation. But such references should not be taken to be more definite than they are. There are, for example, many ways of trying to implement the basic ideas behind "the" many-worlds interpretation. And these do not amount to mere stylistic variants: Each of them gives rise to a very different conception of quantum mechanics.

This is true also of "the" Copenhagen interpretation. Bohr, Heisenberg, and Pauli each held significantly different views on how quantum mechanics should be understood. And the views of von Neumann and of Wigner diverge even more radically from these, although they consider themselves to be proponents of the very same interpretation, and are often taken at their word. Textbook writers typically pay lip service

[1]This passage is quoted from the edited transcript of a radio interview, and appears on page 71 of Davies and Brown (1986).

to the ideas of Bohr, but neither state these clearly and accurately, nor use them to develop any coherent understanding of quantum mechanics.[2] One might still wonder whether there is *some* version of "the" Copenhagen interpretation which is preferable to any other, and clearly superior to all non-Copenhagen views. If so, it would lay claim to be Peierls's one true interpretation of quantum mechanics. Though I do not believe that there is any such version, justifying this belief would require an exhaustive study of the many versions that have actually been proposed, as well as a number that could have been proposed. It is not the purpose of this monograph to undertake such a study. Instead, in the second section of this introduction I shall sketch the basic ideas behind two common versions of the Copenhagen interpretation, and outline what I take to be the chief reasons for rejecting them. This is not intended as a definitive refutation of "the" Copenhagen interpretation, but as a preliminary statement of reasons for looking beyond it.

I am not alone in my dissatisfaction with what I call the Copenhagen interpretation. Although many physicists at least pay lip service to this interpretation of quantum mechanics, there is a significant, and perhaps growing, minority who reject it in favor of something else. Rivals to Copenhagen orthodoxy now include naive realism,[3] (nonlocal) hidden variable theories,[4] the Everett, or many-worlds, interpretation,[5] and the quantum logical interpretation.[6] It is therefore appropriate for me to address the question of the relation of the interpretation to be presented in this mono-

---

[2]David Bohm's (1951) *Quantum Theory* is a welcome exception to this generalization.
[3]See Ballentine (1970) for one physicist's presentation of this view. It is arguable that Einstein held a naive realist view of quantum mechanics. I have criticized this view in Healey (1979).
[4]See, for example, Vigier (1982).
[5]For defenses of this view see, for example, DeWitt and Graham (1973), and Geroch (1984). For objections, see, for example, Healey (1984), and Stein (1984).
[6]For one physicist's presentation of this view see Finkelstein (1962). The quantum logical interpretation seems more popular among philosophers and mathematicians: see Putnam (1968), Bub (1974), and Friedman and Putnam (1978).

graph to these other "unorthodox" approaches. It is clear that my presentation should not be considered part of either the naive realist or the quantum logical traditions. I should myself also resist its assimilation into either the hidden variable or the many-worlds tradition; though others may classify the view differently. The important point is that, in my opinion, there are powerful arguments against the usual versions of all the familiar unorthodox interpretations [though these are not so powerful as Peierls implies in the quoted passage; a quantum logical interpretation might be accused of (classical) inconsistency, but I doubt that any of the other interpretations can be shown to be *inconsistent*]. I shall sketch some of these usual versions in the second section, and then argue against them. Again, I must stress that I do not take any of these arguments to provide a definitive refutation of the interpretation against which it is offered. That would at least require considerable sympathetic reconstruction of each interpretation, combined with systematic and wide-ranging criticism. My purpose is the more limited one of establishing a prima facie case against each interpretation to motivate my own presentation of still another interpretation in the rest of this monograph.

The preceding discussion assumes that it is clear both what an interpretation of quantum mechanics would be and that it is necessary to find an interpretation that is in some sense acceptable. In fact, this is not so clear as to go without comment. Let me start from Peierls's equation of an interpretation of a theory with a way of understanding that theory. Why should it be necessary to provide, in addition to the theory itself, something further, namely a way of understanding the theory?

This would be necessary if one took the theory to be an uninterpreted formal system, on the positivist model for a scientific theory.[7] In this model, the theory would require

<hr>

[7]See, for example, Carnap (1939).

supplementation by semantic principles in order for its constituent terms and sentences to be endowed with meaning, and there may be controversy as to just what these semantic principles should be. But the general positivist model for a scientific theory has justifiably come under sustained criticism in recent philosophy of science, and there seems little reason to suppose that quantum mechanics conforms to this model better than other theories. Furthermore, disputes about how quantum mechanics should be understood extend down to the level of disagreement over how one would go about formalizing the theory; consequently, there is no agreed formal system whose semantics is in doubt.

It is true that there is widespread agreement that quantum mechanics employs certain by now well-understood mathematical structures: For example, in quantum mechanics dynamical variables are representable by self-adjoint operators on a Hilbert space, whose spectra represent possible values of these variables. But such agreement does not extend to the exact wording of the fundamental principles of the theory, nor even to just what these principles are. For example, the status of the projection postulate – von Neumann's process 1 – has long been highly controversial. There are those who have held some form of this principle to be an essential postulate of the theory; others have taken it to hold only in certain special circumstances; still others have considered the principle to be actually inconsistent with the fundamental principles of quantum mechanics. The exact statement of the basic (Born) probability rules has also been a highly controversial matter: Do these specify probabilities that a quantity has a certain value, that it will or would acquire a certain value on measurement, or that a measuring apparatus will or would record a corresponding result if the quantity is measured? In addition to disagreement over the status and exact formulation of such important theoretical principles, there has been no clearly agreed upon understanding of central no-

5

tions of quantum mechanics such as those of *measurement* and of *quantum states*. Here we have an interpretative problem that more closely fits the positivist paradigm: Just what is meant by terms like 'measurement' and 'quantum state' as they figure in quantum mechanics? Finally, it is well known that the conceptual foundations of quantum mechanics have been plagued by a number of "paradoxes," or conceptual puzzles, which have attracted a host of mutually incompatible attempted resolutions – such as that presented by Schrödinger (1935), popularly known as the paradox of Schrödinger's cat, and the EPR "paradox," named after the last initials of its authors, Einstein, Podolsky, and Rosen (1935). A satisfactory interpretation of quantum mechanics would involve several things. It would provide a way of understanding the central notions of the theory which permits a clear and exact statement of its key principles. It would include a demonstration that, with this understanding, quantum mechanics is a consistent, empirically adequate, and explanatorily powerful theory. And it would give a convincing and natural resolution of the "paradoxes." I should like to add a further constraint: that a satisfactory interpretation of quantum mechanics should make it clear what the world would be like if quantum mechanics were true. But this further constraint would not be neutral between different attempted interpretations. There are those, particularly in the Copenhagen tradition, who would reject this further constraint on the grounds that, in their view, quantum mechanics should not be taken to describe (microscopic) reality, but only our intersubjectively communicable experimental observations of it. It would therefore be inappropriate to criticize a proposed interpretation solely on the grounds that it does not meet this last constraint. But this constraint will certainly appeal to philosophical realists, and for them at least it should count in favor of an interpretation if it meets this constraint, as does

6

the interpretation presented in this monograph – or so I hope to show.[8]

Granted that quantum mechanics requires some interpretation, why is a *new* interpretation needed? Why are none of the interpretations so far offered acceptable? Even a preliminary answer to these questions calls for a discussion of the key points of some of the more prominent contemporary approaches to quantum mechanics. It is convenient to introduce this discussion by referring to the Born rules, which are customarily taken to be the central probabilistic principles of quantum mechanics: Different approaches may be distinguished by their different interpretations of the Born rules. Leaving aside joint probabilities, these may be taken to have the following form:

$$\text{prob}_{\psi}(\mathcal{A} \in \Omega) = p. \qquad (1.1)$$

Here p is a real number between zero and one (including those limits), $\mathcal{A}$ is a quantum dynamical variable, Ω is a (Borel) set of real numbers, and ψ is a mathematical representative of an instantaneous quantum state. A preliminary reading of (1.1) is as follows: "In quantum state ψ, the probability of finding that the value of $\mathcal{A}$ lies in Ω is p." But how is the phrase 'of finding' to be understood? Is this phrase just a redundant rhetorical device inserted to draw attention to the fact that instances of (1.1) are testable by repeatedly measuring the value of $\mathcal{A}$ on each of a large number of similar systems in quantum state ψ and observing in what fraction of the tested cases that value lies in Ω? Or would the omission of this phrase constitute a substantive distortion of the content

---

[8]It is interesting to note that its appeal extends also to certain antirealists. Before offering his own interpretation of quantum mechanics, Van Fraassen (1981), certainly no scientific realist, formulates the interpretative task of the philosopher of science as that of "describing how the world can be the way that scientific theories say that it is" (p. 230).

of (1.1), which is intended to apply explicitly to the results of *measurements* of $\mathcal{A}$, and not to the value $\mathcal{A}$ has independent of whether or not it is measured?

One main approach to quantum mechanics takes the first option: According to an approach I have elsewhere characterized as **naive realist** (see Healey, 1979), the Born rules apply directly to possessed values of quantities, and only derivatively to results of measurements of these quantities. According to naive realists every quantum dynamical variable always has a precise real value on any quantum system to which it pertains, and the Born rules simply state the probability for that value to lie in any given interval. Thus, for them, the Born rules assign probabilities to events involving a quantum system σ of the form "The value of $\mathcal{A}$ on σ lies in Ω." A properly conducted measurement of the value of $\mathcal{A}$ on σ would find that value in Ω just in case the value actually lies in Ω (or, at least, would have lain in Ω had the measurement not altered the value of $\mathcal{A}$ while measuring it, just as a thermometer might alter the temperature of a substance while taking it, an effect which in this case may be corrected for to yield the hypothetical undisturbed temperature of the substance).

Perhaps the main problem for the naive realist comes from a set of arguments based on no-hidden-variable proofs.[9] These seem to show that even if the precise values principle endorsed by the naive realist were true, it would be impossible to assign a value to every dynamical variable on each of a large number of similar systems in such a way that for each quantity, the fraction having each value is even close to the probability specified by the Born rules. It seems mathematically impossible to interpret the Born rules uniformly as

[9]See, for example, Healey (1979), and Redhead (1987). The main no-hidden-variable results are contained in Gleason (1957) and Kochen and Specker (1967). The naive realist approach also has particular difficulties in accounting for violations of the Bell inequalities without postulating a kind of instantaneous action at a distance which is in conflict with the basic principles of relativity theory; see Chapter 5.

8

giving probability distributions over possessed values of dynamical variables. Since this claim continues to be disputed, and the arguments surrounding it are both complex and dependent on highly mathematical results, I cannot pursue the issue further here. But the naive realist approach has always been at most an interesting heretical alternative to the more orthodox Copenhagen viewpoint which I consider next.

In the Copenhagen view, the Born rules explicitly concern the probabilities for various possible *measurement results*. They do not concern possessed values of dynamical variables. Indeed, according to this view, on each system there will always be some dynamical variables which do not possess precise values. In the Copenhagen interpretation, the Born rules assign probabilities to events of the form "The measured value of $\mathscr{A}$ on σ lies in Ω." Since the statement of the Born rules then involves explicit reference to measurement (or observation), to complete the interpretation it is necessary to say what constitutes a measurement. Proponents of the Copenhagen interpretation have typically either treated 'measurement' (or 'observation') or cognates as primitive terms in quantum mechanics, or else have taken each to refer vaguely to "suitable" interactions involving a "classical system."

Each of these accounts is problematic. If "measurement" remains a primitive term, then it is natural to interpret it epistemologically as referring to an act of some observer which, if successful, gives him or her knowledge of some structural feature of a phenomenon. But then, quantum mechanics seems reduced to a tool for predicting what is likely to be observed in certain (not very precisely specified) circumstances, with nothing to say about the events in the world which are responsible for the results of those observations we make, and with no interesting implications for a world without observers. And indeed this instrumentalist/pragmatist conception of quantum mechanics has often gone along with the Copenhagen interpretation. On the other hand, if a measurement is a "suitable" interaction with a

9

"classical system," we need to know what interactions are suitable, and how there can be any "classical systems," if quantum mechanics is incompatible with and supersedes classical mechanics.

In order to clarify and amplify these problems, it is useful to distinguish between two different versions of the Copenhagen interpretation. I suspect that whereas the first version is more familiar to many physicists, it is the second version which comes closer to representing Bohr's own view. In what I shall call the *weak version* of the Copenhagen interpretation, the dynamical properties of an individual quantum system are fully specified by means of its quantum state. A dynamical variable $\mathcal{A}$ possesses a precise real value a_i on a system if and only if that system is describable by a quantum state for which the Born rules assign probability one to the value a_i of $\mathcal{A}$. In that state, a measurement of $\mathcal{A}$ would certainly yield the value a_i. In other states, for which there is some chance that value a_i would result if $\mathcal{A}$ were measured, and some chance that it would not, it is denied that $\mathcal{A}$ has any precise value prior to an actual measurement of it. Nevertheless, within the limits of experimental accuracy, measurement of a dynamical variable always yields a precise real value as its result, and this raises the question of the significance to be attributed to this value, given that it is typically not the value the variable possessed just before the measurement, nor the value it would have had if no measurement had taken place. One natural response is to say that the measured variable *acquires* the measured value as a result of the measurement: And then the Born rules explicitly concern the probabilities that dynamical variables acquire certain values upon measurement. Putting this response together with the condition for ascribing a precise real value to a variable given earlier, one concludes that after a precise measurement of a dynamical variable, a system is describable by a quantum state for which the Born rules assign probability one to the measured value of that variable. And this is one form of the **projection postulate** – a controversial principle which purports to specify how the quantum state of a system changes as a result of the measurement of a dynamical variable on that system.

In this popular version of the Copenhagen interpretation, a quantum system has a dynamical property if and only if it is describable by a quantum state for which the Born rules assign probability one to that property: A system has a property just in case observation would be certain to reveal that property. And for each individual quantum system there is a unique quantum state which in this way completely specifies the dynamical properties of that system. It follows that conjugate quantities such as (corresponding components of) position and momentum never have simultaneous precise values on any quantum system. The Heisenberg indeterminacy relations are then interpreted as limiting the precision with which (say) position and momentum are simultaneously defined on any individual quantum system.

Since, in this version, measurement effects significant changes in the dynamical properties of a system, it is important for a proponent of the interpretation to specify in just what circumstances such changes occur. One might expect that such a specification would be forthcoming in purely quantum mechanical terms, through a quantum mechanical account of measuring interactions. Such an account would show how a physical interaction between one quantum system and another, which proceeds wholly in accordance with the principles of quantum mechanics, can effect a correlation between an initial value of the measured variable on one system (the "object" system) and a final "recording property" on the other ("apparatus") system. The problem of giving such an account has become known as the **quantum measurement problem**. A solution to the measurement problem would explain the reference to measurement in the Born rules in purely physical (quantum mechanical) terms, and would also show to what extent the projection "postulate" may be considered a valid principle of quantum mechanics. Unfor-

11

tunately, despite much effort devoted to this task, it seems that such an account *cannot* be given within the constraints of this version of the Copenhagen interpretation: The measurement problem has proven intractable. Again, this is a complex and technical area which cannot be adequately reviewed here.[10] But the key difficulty may be stated quite simply. It is that many initial states of an object system give rise to final compound object+apparatus quantum states which, in the present interpretation, imply that the apparatus fails to register any result at all (its pointer points nowhere at all)! For, in such a final compound quantum state, the Born rules do not assign probability one to any recording property of the apparatus system.

The "paradox" of Schrödinger's cat provides a dramatic illustration of this difficulty. As is well known, Schrödinger (1935) described a thought experiment in which the state of a microscopic device (a radioactive nucleus) is coupled to a macroscopic system in such a way that if the nucleus has definitely decayed after a certain period of time a cat will definitely be dead, whereas if the nucleus has definitely not decayed after this time period, the cat will definitely be alive. The cat may be thought of as a rather unconventional (and indeed ethically questionable!) apparatus, whose condition records whether or not the nucleus has decayed. If the state of the nucleus at the beginning of the period is such that there is a 50 percent chance of it being observed to decay during the period, then, in the absence of any observation or other intervening interaction, at the end of the period the quantum state of the joint system nucleus+cat (+other coupled intermediate systems) will correspond to the cat's being *neither* dead nor alive: though, of course, there would then be a 50

---

[10]The classic presentation of the measurement problem is by Wigner (1963). That a more general quantum mechanical treatment fails to solve the problem is indicated, for example, by Fine (1970), whose proof is supplemented by Shimony (1974).

percent chance of *observing* the cat to be dead (an unusual variant on the maxim that it was curiosity that killed the cat!).

The popular version of the Copenhagen interpretation faces the measurement problem because it takes each individual quantum system (for example, a single silver atom) to be describable by a quantum state which wholly determines its dynamical properties. But there is another version of the Copenhagen interpretation which views quantum states differently. According to what I shall call the *strong version* of the Copenhagen interpretation, it is mistaken to ascribe a quantum state to an individual system; a quantum state may be correctly ascribed only to an "ensemble" – that is, to a set of similar systems (such as a beam of silver atoms emerging from an oven), which share a certain physical history not possessed by a random collection of spatiotemporally dispersed similar systems. In this view, to ascribe a quantum state to an ensemble at a time is to say nothing about the dynamical properties of its elements at that time: Rather, the ascription of a quantum state to an ensemble is just a theoretical device which permits (correct) predictions concerning the statistics of experimental results, if the dynamical properties of members of the ensemble are observed. In this sense, the Born rules (together with techniques for describing particular ensembles by particular quantum states) exhaust the significance of the quantum state. Quantum mechanics simply has nothing to say about the dynamical properties of any quantum system at a time when it is not being observed.[11]

The weak version of the Copenhagen interpretation appeared to imply that a measuring apparatus may end up in a state in which it fails to have any well-defined property corresponding, for example, to the pointer position. And this implication conflicts with the belief we normally take to be

---

[11]Indeed, in Bohr's own view, it would be *meaningless* to ascribe any dynamical property to a quantum system except in the context of a well-defined experimental arrangement suitable for the observation of that property.

warranted by our everyday observations, that no macroscopic apparatus ever fails to have *some* such directly perceptible property. However, in the strong version of the interpretation, one cannot apply quantum mechanics to an individual measurement interaction at all, but at most to an ensemble of similar such interactions. And even then, if one were to describe the ensemble of compound object + apparatus systems quantum mechanically, it would be illicit to infer anything about the final condition of the apparatus systems from the postinteraction quantum state of the ensemble. The restricted significance of the quantum state permits no such inference. The problem with the strong version does not arise because in certain circumstances a description in terms of the quantum state contradicts something we take ourselves to be justified in believing about the condition of a macroscopic apparatus after a measurement. It arises because this description never implies anything at all about that condition. On the one hand, the strong version of the Copenhagen interpretation takes the Born rules to yield probabilities for the possible results of measurements; whereas, on the other hand, in this view, a direct application of quantum mechanics to the compound object + apparatus systems involved in a measurement interaction fails to show that a measurement will yield any result at all.

The view tries to solve this problem by requiring that the measuring apparatus be described *classically*. But what does this mean, and how does it help? It is plausible to suppose that to treat the measuring apparatus classically is to assume that it possesses a classical dynamical state, specified by a point in a classical phase space, and, perhaps, also that this state evolves in accordance with the laws of classical dynamics during the measurement interaction. That supposition would, of course, be highly problematic. Even though it may be macroscopic, the apparatus presumably consists of atoms; and it is the empirical inadequacy of classical mechanics as applied to atomic systems which necessitated the transition to quantum mechanics.

I suggest therefore that this interpretative supposition is wrong, and that, on the strong Copenhagen view, to give a classical description of the apparatus is not to assume that classical mechanics holds for it. Rather, it is to assume that the apparatus records a definite outcome of the measurement – that is, after the measuring interaction, the apparatus may be assigned a dynamical state which is classical in the minimal sense that the apparatus definitely has some enduring macroscopic property which can serve to record the outcome of the interaction. On the strong Copenhagen view, one cannot prove that the apparatus records a definite outcome from any description of the apparatus by means of a quantum state. But in the strong version of the Copenhagen interpretation (unlike the weak version) a definite measurement outcome is not *inconsistent* with any such description. And since no such inconsistency is threatened, the measurement problem does not arise in its usual form. Schrödinger's cat is definitely alive or dead before it is observed, but quantum mechanics cannot say which. Indeed, in this view, to apply quantum mechanics to the cat, considered as a measuring apparatus, one must simply *assume* that it ends up either alive or dead at the conclusion of the interaction, so that one cannot even show that it will end up one or the other.

Why can we not just take this as an independent assumption, which is amply justified by our everyday observations of macroscopic objects such as cats? There would be no prima facie inconsistency involved in incorporating *some* such assumption into the strong version of the Copenhagen interpretation. But exactly what form could that assumption take? The problem is that it proves very difficult to give a precise general statement of such an assumption consistent with the rest of the interpretation. The assumption that all systems always have precise values for all dynamical variables would contravene the Copenhagen understanding of the Heisenberg relations, which states that these relations impose limits on the precision with which conjugate variables are simultaneously defined on any system:[12] it would also lead to many of the same problems faced by a naive realist interpretation. On the other hand, to give content to the weaker assumption that a measuring apparatus always emerges from a measurement interaction with some definite recording property at least requires a precise specification of what is to count as a measuring apparatus, a measurement interaction, and a recording property. But no such specification is offered.

Moreover, one cannot simply assume that experimental physicists have developed clear operational tests for these notions, which might then be taken to provide at least a sort of working understanding of them. For if one considers the way in which quantum mechanics is applied, it becomes apparent that there is no unique correct way of dividing up any given experimental arrangement into object system and apparatus system. Many such divisions are possible, even though the results depend in detail on just how the division is made. All one can do is to apply a rule like that stated by Bell (1987):

.... the following rule..., although ambiguous in principle, is sufficiently unambiguous for practical purposes: *put sufficiently much into the [object] system that the inclusion of more would not significantly alter practical predictions.* (p. 124)

We have now pushed the problem for the strong version of the Copenhagen interpretation back until it assumes the form in which Bell states it:

The problem is this: quantum mechanics is fundamentally about "observations". It necessarily divides the world into two parts, a

---

[12]Bohr himself would reject this assumption for the stronger reason that it countenances ascription of dynamical properties to quantum systems outside the context of any experimental arrangement suitable for their observation.

part which is observed, and a part which does the observing. The results depend in detail on just how this division is made, but no definite prescription for it is given. All we have is a recipe which, because of practical human limitations, is sufficiently unambiguous for practical purposes. So we may ask with Stapp: "How can a theory which is *fundamentally* a procedure by which gross macroscopic creatures, such as human beings, calculate predicted probabilities of what they will observe under macroscopically specified circumstances ever be claimed to be a complete description of physical reality?" (ibid., p. 124)

Now Stapp's question (echoed by Bell in the last sentence of the quoted passage) actually raises a further problem for the Copenhagen interpretation, in either version. The problem arises as follows. Quantum mechanics is clearly a very successful theory with very wide applications. Moreover, it is commonly held to be not only predictively accurate, but also explanatorily powerful – perhaps even *the* most explanatorily powerful theory of contemporary physics. But if the Copenhagen interpretation were right, then, strictly, all that quantum mechanics describes, and permits one to predict, are the results of observations. Now we ordinarily assume that there is far more going on in the world than we are able to observe, and that the primary explanatory task of physics is to account for what actually occurs, whether or not we happen to observe it. Indeed, we ordinarily assume that a satisfactory explanation even of what we do observe will appeal to those hidden processes and mechanisms which are responsible for our observations.

These assumptions support the charge that the Copenhagen interpretation of quantum mechanics is both descriptively and explanatorily incomplete. First, accepting this interpretation means accepting that quantum mechanics is descriptively incomplete on a massive scale. It does not describe the structure of elementary particles, nuclei, atoms, metals, superfluids, superconductors, lasers, crystalline solids, and semiconductors: It merely describes what we find when we observe such things. Indeed, on this view, quantum mechanics would have nothing to say about a world in which no observations ever took place; and depending on exactly what is taken to be required for an observation, it may not even have anything to say about our world before there were entities such as humans capable of observing it. Second, the Copenhagen interpretation renders quantum mechanics explanatorily defective. Not only does it fail to explain phenomena that are not observed – including phenomena which we ordinarily take ourselves to have excellent reasons to suppose in fact occur, such as thermonuclear reactions in distant stars – but also the explanations it does offer of those phenomena that are observed amount to little more than predictions of their occurrence, with no independent accompanying account of whatever (perhaps microscopic) processes and events gave rise to them.

Although this last problem for the Copenhagen interpretation may seem much more profound than those raised earlier, I have left it until last because proponents of the Copenhagen interpretation have a ready response to this problem, which draws heavily on certain general views of the nature and purposes of science, and the character of scientific explanation. The basic thrust of the response is to argue that the goal of science is to develop theories and techniques for successfully predicting our observations, that talk of any "reality" independent of our observations is metaphysical and unscientific, and consequently it is not part of science to describe such a "reality," and that, since explaining a phenomenon differs only psychologically from describing it as conforming to general regularities (insofar as we like an explanation to provide a "mental picture" of the phenomenon to be explained), we are bound to find quantum mechanical "explanations" unsatisfying just because they exclude the possibility of forming any "mental picture" of the phenomena.

This response to the charge that the Copenhagen interpre-

18

tation would render quantum mechanics descriptively and explanatorily incomplete clearly rests on particular answers to central, though disputed, questions in the philosophy of science. An adequate rebuttal of the response would therefore require an extensive philosophical critique of these answers – something which has no place in this monograph. Therefore, I shall simply state my contrary convictions: that it is a primary goal of science to explain phenomena in the natural world; that our observations are only of interest insofar as they give us access to some of these phenomena; and that there are objective criteria for assessing the adequacy of a scientific explanation which have nothing to do with a psychological preference for mentally picturable models. These convictions are a primary motivation for the project I undertake in this book. Moreover, I suspect that it was similar convictions which motivated the development of quantum mechanics, and which continue to play a significant role in influencing people to devote themselves to those branches of contemporary physics which rest on this theory. It would be disappointing if the only way to understand much of contemporary physics were to reject beliefs which motivated one to make the attempt to do so.

I turn now to a brief critical discussion of three unorthodox approaches to quantum mechanics: Each of these is motivated by dissatisfaction with the perceived incompleteness of the Copenhagen interpretation, and also by the perceived inadequacy of its treatment of the measurement process. Everett (1957) proposed a radical reinterpretation of quantum mechanics, and his proposal gave rise to what has become known as the many-worlds interpretation, though his own name ('relative state formulation') may in the end prove less misleading. According to this interpretation, all interactions, including measurement interactions, may be considered as quantum mechanical interactions internal to some compound system; there is no need to assume the existence of any classical systems; and the entire universe may itself be treated as

19

a single immensely complex quantum system, and assigned its own quantum state. Every other quantum system is treated as a subsystem of the universe. Interactions among such subsystems would soon prevent each from being described by its own unique quantum state: Nevertheless, each subsystem may always be assigned many *relative (quantum) states*, in such a way that there is a correlation between the relative states of any two systems which together compose the universe. Moreover, the dynamical properties of a subsystem in a particular relative state are just those which would be assigned probability one by application of the Born rules to that state. In a model universe composed just of an "object" system and an "apparatus" system, a quantum mechanically described measurement interaction would give rise to a quantum state for the compound system for which the Born rules do not assign probability one to any property corresponding to a definite measurement outcome. Nevertheless, each of the many apparatus *relative* states does assign probability one to some definite measurement outcome – though typically a *different* outcome for each such state. It is as though the pointer points everywhere at once, or Schrödinger's cat is both dead and alive!

This raises two problems, one internal and one empirical. The internal problem is to show why it is not straightforwardly inconsistent to maintain that an "apparatus" system simultaneously possesses incompatible properties. One version of the many-worlds interpretation seeks to restore consistency by maintaining that measurement interactions induce a fission of the systems which undergo them: Before the interaction there is a single apparatus system, but after the interaction there are many, recording different results. This seems both ontologically extravagant, and observationally unwarranted, since we never observe such fissions. Instead, we always observe a single, definite result of any properly conducted quantum measurement, even though in the many-

20

worlds interpretation there are often supposed to be many different results: This is the empirical problem. Proponents of the interpretation have made strenuous and ingenious efforts to solve these problems, but in my opinion without success.[13] However, despite its problems, the many-worlds interpretation is not without value, for its elegant treatment of many conceptual difficulties of quantum mechanics may provide models for the development of parallel solutions within an alternative interpretation like that offered in this book.

Quantum logic is a confusing subject, and it is hard to pin down what is supposed to be involved in various attempts to give a quantum logical interpretation of quantum mechanics. Central to the interpretations of Putnam (1968) and Bub (1974) was the claim that quantum logic permits one to maintain the naive realist's understanding of dynamical states and their relation to the Born rules. Specifically, they held that the adoption of a nonclassical logic in the quantum domain enables one consistently to maintain that every quantum dynamical variable always has a precise real value, and that measurement is merely our way of getting to know that value, in the face of the arguments against this based on the no-hidden-variable proofs mentioned in footnote 9. Apart from general objections to the adoption of a nonclassical logic for empirical reasons,[14] there are more specific objections to this particular proposal, viewed as an attempt to breathe new life into naive realism. The *content* of the claim that every dynamical variable has a precise value is quite obscure once one has adopted quantum logic, given that the standard (classical) inferences can no longer be drawn from it. And, in particular, it is by no means clear that the truth of this claim suffices to permit a naive realist reading of the Born rules,

[13] See the references given in footnote 5, this chapter.
[14] See, for example, Dummett (1978), as well as Kripke's unpublished talk, "The Question of Logic," given at the University of Pittsburgh in 1974.

according to which the object system possessed the specific value revealed by a measurement prior to, or independent of, the occurrence of that measurement.[15]

Finally, I wish to briefly consider hidden variable theories. A clear motivation behind the construction of such theories has been the belief that some more complete account of microscopic processes is required than that provided by quantum mechanics according to the Copenhagen interpretation. The general idea has been to construct such an account by introducing additional quantities, over and above the usual quantum dynamical variables (such as de Broglie's pilot wave, Bohm's quantum potential, or fluctuations in Vigier's random ether), and additional dynamical laws governing these quantities and their coupling to the usual quantum variables. The primary object is to permit the construction of a detailed dynamical history of each individual quantum system which would underlie the statistical predictions of quantum mechanics concerning measurement results. Though it would be consistent with this aim for such dynamical histories to conform only to indeterministic laws, it has often been thought preferable to consider in the first instance deterministic hidden variable theories. A deterministic hidden variable theory would underlie the statistical predictions of quantum mechanics much as classical mechanics underlies the predictions of classical statistical mechanics. In both cases, the results of the statistical theory would be recoverable after averaging over ensembles of individual systems, provided that these ensembles are sufficiently "typical": but the statistical theory would give demonstrably incorrect predictions for certain "atypical" ensembles.

Now, as Bell (1964) first showed, no deterministic hidden variable theory can reproduce the predictions of quantum mechanics for certain composite systems without violating a

[15]For objections along these lines, see Putnam (1981), Stairs (1983), and Healey (1977).

principle of *locality*. And this principle seems firmly grounded in basic assumptions concerning the lack of physical connection between spatially distant components of such systems; and the impossibility of there being any such connection with the property that a change in the vicinity of one component should instantaneously produce a change in the behavior of the other. Further work attempting to extend Bell's result to apply to indeterministic hidden variable theories has shown that there may be a small loophole still open for the construction of such a theory compatible with the relativistic requirement that no event affects other events outside of its future light-cone.[16] But to my knowledge no plausible theory has succeeded in exploiting this loophole, and I shall not seek to do so in this book.

Existing hidden variable theories, such as that of Vigier (1982), are explicitly nonlocal, and do involve superluminal propagation of causal influence on individual quantum systems, although it is held that exploiting such influence to transmit information superluminally would be extremely difficult, if not actually impossible. Against this, it has often been maintained that any superluminal transmission of causal signals would be explicitly inconsistent with relativity theory: If this were so, such nonlocal hidden variable theories could be immediately rejected on this ground alone. But as I shall argue in Chapter 5, relativity does not explicitly forbid such transmission. Nonlocal hidden variable theories like that of Vigier can conform to the letter of relativity by introducing a preferred frame, that of the "subquantum ether," with respect to which superluminal propagation is taken to occur. By doing so they avoid the generation of so-called causal paradoxes. But they also thereby violate the spirit of relativity theory by reintroducing just the sort of privileged reference frame which it was Einstein's great achievement to have shown to be unnecessary for the formulation of (classical)

[16] See especially Jarrett (1984).

23

mechanics and electromagnetism. Moreover, they do so while at the same time giving reasons as to why it would be at least extremely difficult to find out what this frame is. And this is why I find these theories so implausible. The principle that a fundamental theory can be given a relativistically invariant formulation seems so fundamental to contemporary physics that no acceptable interpretation of quantum mechanics should violate it.[17]

There is a more basic reason why I cannot accept a hidden variable theory as an interpretation of quantum mechanics. A hidden variable theory is, fundamentally, a separate and distinct theory from quantum mechanics. To offer such a theory is not to present an interpretation of quantum mechanics but to change the subject.[18] Now it does follow from my characterization of what is involved in interpreting a theory that, prior to interpretation, the bounds of a theory are somewhat ill-defined. And my own interpretation will draw the bounds of quantum mechanics in a perhaps unusual way. But there seem to be clear reasons for denying that a hidden variable theory should count as part of quantum mechanics. One reason is that a hidden variable theory incorporates quantities additional to the quantum dynamical variables. Another is that hidden variable theories are held to underlie quantum mechanics in a way similar to that in which classical mechanics underlies the *distinct* theory of statistical mechanics. A final reason is that a hidden variable theory (at least typically) is held to be empirically equivalent to quantum me-

---

[17] Of course, an interpretation of nonrelativistic quantum mechanics must portray that theory as violating relativistic invariance. But there should be nothing inherent in the interpretation itself which prevents its generalization to a fully relativistically invariant quantum mechanics.

[18] This point may have been obscured by the approach, and especially by the title, of Bohm (1952a, 1952b). In these papers Bohm combined objections to the Copenhagen interpretation with the proposal of a hidden variable theory to underlie quantum mechanics, now interpreted according to his own avowedly non-Copenhagen approach. But his non-Copenhagen approach to quantum mechanics did not require the proposal of a hidden variable theory: It merely *permitted* this.

chanics only with respect to a restricted range of conceivable experiments, while leading to conflicting predictions concerning a range of possible further experiments which may, indeed, be extremely hard to actualize.

1

Overview

No interpretation of quantum mechanics can be wholly comprehended or validated before carefully investigating all of its details. I begin such an investigation of the present interpretation in the next chapter. But some of these details will prove rather intricate. The reader may need help in seeing the forest formed by the trees, and the skeptical reader may require encouragement before he or she agrees to enter its thickets. In this first chapter I therefore present a preliminary account of the basic ideas of the interpretation to be given, which may serve as a guide to what is to follow. Just as parts of a travel guide may seem puzzling before one actually visits the region described, so also parts of this chapter will likely seem much clearer when they are reread in the light of the rest of the book.

There is another reason for beginning with a general overview. The present interpretation arose as what seemed to be the most promising way of implementing a certain general approach to quantum theory. But, though I am confident of the worth of this approach, the present interpretation may well come to be regarded as only a fruitful first attempt at its detailed implementation, requiring more or less extensive technical revision if it is to prove ultimately satisfactory. If so, this overview may continue to be useful as a guide to an interpretation whose technical framework differs more or less radically from that underlying the present interpretation.

Let me start with the concept of the *state* of a system. In quantum mechanics this is normally taken to be represented by a mathematical object, such as a wave function, state vector, or density operator. Although this remains broadly true in the present approach, it becomes important to distinguish between two different conceptions of the state of a system: the **dynamical state** and the **quantum state**. In most interpretations, including the present one, the primary role of the quantum state is to generate probabilities, concerning the possible outcomes of measurements, via the Born rules. Thus, if a system is ascribed quantum state ψ, then the probability that a measurement of dynamical variable $\mathcal{A}$ would give a value lying in (Borel set) Ω is written as prob_ψ $(\mathcal{A} \in \Omega)$, where this probability is calculated according to the appropriate quantum algorithm [for example, prob_ψ $(\mathcal{A} \in \Omega) = (\psi, \mathbf{P}^{\mathcal{A}}(\Omega)\psi)$, where ψ is the system's state vector, and $\mathbf{P}^{\mathcal{A}}(\Omega)$ is the projection operator corresponding to the property $\mathcal{A} \in \Omega$]. On the present interpretation, a quantum state may be legitimately ascribed to a single quantum system (and not just to a large ensemble of similar systems), but only in certain circumstances: A system does not always have a quantum state. The circumstances in which it is legitimate to ascribe a quantum state to a system will be described below. For present purposes it is necessary merely to note that these circumstances are not universal. Nevertheless, every quantum system *always* has a dynamical state. Consequently, there can be no general identification between a system's quantum state and its dynamical state; nor is it always true that one determines the other.

The dynamical state of a system at an instant may be identified with a truth-value assignment to all sentences ascribing a quantum dynamical property to that system at that instant. Each such sentence may be written in the form $\mathcal{A} \in \Omega$, where $\mathcal{A}$ stands for a quantum dynamical variable (such as x-component of position, or total spin), and Ω for a (Borel) set of real numbers. The truth-value assignment is such as to assign either true or false to each such **elementary sentence**.

27

It will also satisfy certain restrictions, including restrictions designed to mirror the intended semantics of property ascriptions: For example, if $\Omega \subset \Omega'$ and the sentence $\mathcal{A} \in \Omega$ is true, then so is $\mathcal{A} \in \Omega'$.

Now, in the weak version of the Copenhagen interpretation, the dynamical state of a system is completely specified by the same mathematical object that represents its quantum state. On that interpretation, for each quantum system, at any time, there is a mathematical representative of its quantum state at that time; and this mathematical representative *also* determines its dynamical state, via the Born probability rules. Specifically, the truth-value assignment that constitutes the dynamical state is generated as follows: An elementary sentence is true just in case the Born rules assign it probability 1 in the given quantum state.

By contrast, in the naive realist interpretation, the dynamical state of a system is relatively independent of its quantum state. Though an elementary sentence (concerning a quantity with a discrete range of possible values) is true if it is assigned probability 1 by the quantum state via the Born rules, and false if it is assigned probability 0, the quantum state does not further restrict the dynamical state. Rather, the dynamical state is generated (in a way analogous to classical mechanics) from an assignment of a precise real value to each dynamical variable of the system. An elementary sentence $\mathcal{A} \in \Omega$ is true just in case some sentence $\mathcal{A} = x$ is true, for some real $x \in \Omega$.

In the present approach, a system's dynamical state is not always determined by its quantum state (nor vice versa). But neither is the dynamical state generated (as in the naive realist interpretation) from an assignment of a precise real value to each dynamical variable of the system. Loosely speaking, a system's dynamical state will (typically) turn out to be more definite than it would be in the weak Copenhagen view, but less definite than it would be in the naive realist view.

What, then, determines a system's dynamical state in the present interpretation? Basically, the instantaneous dynamical state of a system is determined by a *set* of mathematical objects (such as state vectors) of the type used to represent the quantum state of a system. This **state set** will include one object, such as a state vector, corresponding to the system in question, but that **system representative** may or may not simultaneously represent the system's quantum state (if any). And the state set will always also include one element corresponding to each quantum subsystem of the given system. For this reason, some of a system's properties are fixed by those of its proper subsystems (if any): I call these the **reducible** properties of the system. But a system generally also has **irreducible** dynamical properties which are not so fixed. Such irreducible properties give rise to a kind of holism which, in the present interpretation, underlies quantum nonseparability (this will be further explored in Chapter 5). Moreover, just as some dynamical properties of a system are fixed by those of its subsystems, so also may other of its dynamical properties be restricted by those of a supersystem. These "downward" restrictions are weaker in two distinct ways. First, they generally fail to restrict the dynamical state of a system to a single possibility (consistent with the dynamical states of its supersystems): The restrictions may typically be met by any of a range of accessible dynamical states. Second, they apply only when a system is not interacting with any disjoint subsystem of the relevant supersystem.

Because of all these correlations among dynamical states, it is fundamental to the present interpretation to picture quantum systems as comprising a hierarchy, ordered by a (nonspatial) composition relation. Such a picture is independently suggested by reflection on other features of quantum mechanics, in particular the symmetry character of quantum states. Thus, an unexcited helium nucleus is a quantum system with a certain symmetry character: It is a boson. This is a consequence of the way it is composed from its subsystems (two protons and two neutrons), and in turn has important implications for the quantum description of collections of helium nuclei: Their overall quantum state must be symmetric under the exchange of any pair of helium nuclei. But in the present interpretation, this picture cannot be thought of as an independent constraint on the quantum description, perhaps added as an afterthought: It is central to that description from the very beginning. The hierarchy of quantum systems, and the constraints it imposes on dynamical states, is one main topic of Chapter 2. An important second topic of that chapter is the evolution of dynamical states.

In this interpretation, the quantum world consists of a hierarchy of quantum systems, each with its own evolving dynamical state. The systems sometimes interact; such interactions are marked by the presence of interaction terms in their joint Hamiltonian. When a system is free from external interactions, its dynamical state changes smoothly in the following sense: The system representative evolves continuously and deterministically. It is this dynamical law which underlies the time-dependent Schrödinger equation, although that equation does not directly govern the dynamical state of any system – rather it describes the time evolution of *quantum* states. Whether or not they constitute measurements, interactions produce more abrupt changes in the dynamical states of the interacting systems: The system representative may undergo one or more indeterministic transitions. Though these transitions are the ultimate source of quantum probabilities, it is important to note that they are not themselves governed by the Born rules. Indeed, under the present interpretation, the detailed dynamics of these indeterministic transitions remain an open question. The Born rules apply to a quantum system only in certain circumstances – in fact, in just those circumstances in which it is legitimate to ascribe a quantum state to the system. But why is it ever necessary to ascribe a quantum state to a system in addition to its dynamical state?

In understanding the relation between the dynamical state and the quantum state in the present interpretation, it will be helpful to make a comparison with classical mechanics. In classical mechanics, the instantaneous state of a system may be specified by means of a point in the system's phase space. This state not only determines all the current dynamical properties of the system (such as its position, momentum, and energy), but it also constrains the future behavior of the system. Indeed, given the equations of motion of the system (including suitable boundary conditions) its future dynamical properties are also wholly determined by its present state. In the present interpretation of quantum mechanics, it is the instantaneous *dynamical* state which specifies all the current dynamical properties of a quantum system; but the dynamical state does *not* generally also suffice to determine the system's future behavior, nor even the probabilities of various possible future behaviors. How a system is disposed to (be likely to) behave in particular future interactions generally depends not only on *its* dynamical state, but also on the dynamical states of systems of which it is a component. The sole function of the quantum state, in the present interpretation, is to characterize a system's current probabilistic dispositions, some of which will be manifested in a future interaction.

A system's quantum state is thus forward-looking: It summarizes all the information about the dynamical state of the system and of systems correlated to it which is relevant in determining the probabilities of the various possible outcomes of an interaction that it is to undergo. It is possible to summarize such information in this way only in special circumstances: The system must be about to undergo an interaction of a certain type (type M), and (at least typically) it must have been subjected to an interaction of another particular type (type P). It is in these circumstances that it is legitimate to ascribe a quantum state to a system – the legitimacy follows from the fact that these are precisely the circumstances in which the Born rules apply. It is natural to describe M-type interactions as *measurements,* and P-type interactions as *preparations*. But it is important to stress that in the present interpretation it is particular features of a quantum interaction itself which determine whether that interaction is of one of these two types. An M-type interaction may take place in the natural course of events, even though no observer has arranged for it to take place, and nobody knows of its occurrence, let alone its outcome. But exactly how are the Born rules to be understood, and why do they apply in just these circumstances?

Here is a preliminary answer to these questions. The set of all possible physical interactions has two distinguished subsets, the M-type interactions and the P-interactions. These subsets may overlap. When a quantum system σ has undergone a P-type interaction, and is about to undergo an M-type interaction with another quantum system α of an appropriate type, σ may be ascribed a quantum state ψ. In these circumstances, system α is naturally described as the *apparatus* system corresponding to the measurement, but once again it is important to stress that it is the purely quantum mechanical features of α and of the M-type interaction with σ which determine whether this description applies: α may not be macroscopic, and it need not be employed by anybody in an attempt to obtain information about σ. Also, in these circumstances it is natural to describe the P-type interaction as *preparing* the quantum state ψ of σ. To say this does not, of course, imply that anyone brought about the P-type interaction. The interaction between σ and α may leave the dynamical state of α unchanged, or it may put α into some new dynamical state out of a particular set of possible final dynamical states, which it is natural to describe as the *outcome* states for this M-type interaction. The cumulative effect of all the indeterministic transitions of dynamical states during the P- and M-type interactions involving σ is such that the probability that the final dynamical state of α corresponds to a particular outcome (given that it does not simply remain unchanged), is correctly given by the Born rules as applied to ψ. Notice that, whereas the Born rules are usually read as specifying probabilities for properties of σ (or for the results of measurements of these properties), in the present interpretation these are to be read as probabilities for outcome states of α. Strictly speaking, therefore, the events to which the Born rules assign probabilities are of the form "α enters the outcome state corresponding to property $\mathcal{P}$ of σ,"and not of the form "σ has property $\mathcal{P}$," nor of the form "σ acquires property $\mathcal{P}$." (As will be shown in Chapter 3, this correspondence between properties of σ and outcome states of α is set up naturally by the form of the M-type interaction between σ and α.) Note that α may enter the outcome state corresponding to property $\mathcal{P}$ of σ even though σ does not have property $\mathcal{P}$ at any time before, during, or after the M-type interaction with α. In such circumstances, the interaction neither reveals nor creates the property recorded in the outcome state of α.

It follows from this understanding of the Born rules that their application presupposes that a system does in fact undergo an M-type interaction, and the Born rules as applied to quantum state ψ can generally be expected to yield the correct probabilities only when that system has been subjected to an appropriate P-type interaction. A P-type interaction prepares the quantum state ψ of σ, not by inducing a physical transition into that state from some other quantum state, but simply by constituting a necessary condition for the correctness of the Born rules as applied to ψ, and thereby helping to render legitimate the ascription of ψ as the quantum state of σ. I call the present interpretation **interactive** because it assigns such an important role to two types of interaction. The Born rules are taken to give probabilities for possible outcomes of an M-type interaction, whereas quantum states are taken to be legitimately ascribable only to systems which are about to undergo an M-type interaction, and typically only to those which have also previously been subjected to a P-type interaction.

The following simplified model may be helpful in understanding the respective roles of dynamical and quantum states in the present interpretation. Consider an experimental arrangement in which a monochromatic beam of neutral spin-½ particles of the same kind is first passed through a Stern-Gerlach magnet oriented in the z-direction. Then a second Stern-Gerlach magnet oriented at right angles to the first, in the x-direction, is placed so as to intercept only particles which (supposedly) emerge from the first magnet deflected (slightly) in the positive z-direction. Finally, a detector is placed behind the second magnet, which is taken to record whether a particle is deflected in the positive or in the negative x-direction after passing through this second magnet. The interaction with the first magnet prepares the quantum state of those particles whose spin in the x-direction is subsequently measured by the second magnet plus detector.

It is important to note that, in the present interpretation, this quantum state preparation does *not* in fact produce two distinct beams emerging from the first magnet, with each particle in exactly one beam: The *dynamical* state of an emerging particle does not specify that its z-position is restricted to one such beam rather than the other. It is therefore not surprising that [as noted, for example, by Wigner (1963)] a different experimental arrangement could, in principle, reveal interference effects by "recombining the beams." Interaction with the first magnet does indeed *affect* the dynamical states of the particles, but its role in preparing their quantum states is not simply to randomly sort each into one of two emerging beams. It is true that the interaction between the particles emerging from the first Stern-Gerlach magnet and the second magnet-plus-detector system does not produce a record for every particle that emerges from the first Stern-Gerlach magnet. And even though (in the present interpretation) those particles that are recorded by the detector need not be in any way distinguished by their prior dynamical histories from those that are not, it is natural to incorrectly assume that it is only the former which constituted an upper beam emerging

34

from the first Stern-Gerlach magnet. After all, classically it is true that a particle would be registered by the detector of the second Stern-Gerlach device only if it had previously been deflected into an upper beam emerging from the first Stern-Gerlach magnet. But this is false, quantum mechanically, at least in the present interpretation.

In what sense, then, does the interaction with the first Stern-Gerlach magnet prepare the quantum state of particles that emerge from it? Ideally, each particle emerging from the first Stern-Gerlach magnet either will be detected as having passed through the second Stern-Gerlach magnet or will not be so detected. This holds even though there is no feature of the dynamical state of such a particle which determines, or is even probabilistically relevant to, whether or not it will be detected. The particles emerging from the first Stern-Gerlach magnet thus constitute two disjoint sets, distinguished *only* by their actual future behavior, and not by their present dynamical states. If a particle will, in fact, subsequently be detected after passing through the second Stern-Gerlach device, then it is legitimate to ascribe to it a quantum state corresponding to z-spin-up when it emerges from the first Stern-Gerlach magnet. The ascription is justified insofar as each conditional probability, for recording a particular x-spin *given* that a particle produces *some* recording of x-spin, is correctly specified by the Born rules as applied to this quantum state.

This example illustrates two important points about the ascription of quantum states. The first point is that one essential factor in determining the present quantum state of a system is a specification of the future M-type interactions that system will actually undergo, including a specification as to whether or not such an interaction will in fact result in a change in the dynamical state of the relevant system α. Without such a specification, it is not strictly legitimate to ascribe any quantum state to a system. It is nevertheless natural and permissible to make such an ascription hypothetically, on the *presupposition* that the system will in fact interact with an M-

type system in such a way as to effect a change in its dynamical state, with no intermediate interactions to undo the effects of any initial *P*-type interaction. And if such a hypothetical ascription is made, the quantum state thereby ascribed does not depend on the *kind* of measurement that might be performed by the supposed future interaction with an *M*-type system. This explains the common assumption that one may ascribe a quantum state to a system irrespective of what measurements may (or may not) be subsequently performed on that system.

The second, and related, point is that even when a specification of the present dynamical state of a system is supplemented by complete information as to the present dynamical states of all other correlated systems, this *still* does not suffice to determine the present quantum state of the given system (or even whether it has any quantum state). In contrast to classical mechanics, no amount of information concerning present dynamical states suffices to determine either what interactions a system will undergo, or whether these will produce an outcome in some "apparatus" system. Consistent with a complete specification of the instantaneous dynamical states of all systems, a given quantum system may be assigned quantum state ψ_1, incompatible quantum state ψ_2, or no quantum state at all. Thus, if a particle emerging from the first Stern-Gerlach magnet with a given dynamical history will be recorded as having passed through the second magnet, it may be assigned a quantum state corresponding to z-spin-up as it emerges from the first magnet. If it will be recorded as having *not* passed through the second magnet (e.g., by placing a detector in the "lower beam" emerging from the first magnet), then it may be assigned a quantum state corresponding to z-spin-down as it emerges from the first magnet. If it will not be recorded in either of these ways, or in any other relevant way, then it may be assigned no quantum state.

Note that quantum state ascriptions are in a sense retrospective, in the present interpretation. One can only be sure of the instantaneous quantum state of a system at a later time, because its present quantum state is not even determined until later. By this acknowledgment, the present approach incorporates insights of Bohr and Wheeler. Bohr (1963) stressed that an unambiguous application of quantum mechanics, involving quantum state ascriptions and their employment in the Born rules, requires a specification of the complete experimental arrangement, including both measuring devices and preparation devices. And Wheeler (1978) noted that in certain "delayed-choice" experiments, this would have the consequence that even after an interaction which supposedly prepares a quantum state is over, the state of a system would not be determinate until it is decided how it will interact with other experimental devices, including measuring apparatus. In the present approach, the instantaneous *dynamical* state of a quantum system is independent of what interactions (if any) it will undergo; but its instantaneous *quantum* state does depend on the nature of such (future) interactions and the role they play in measurements on the system. This latter dependence does not, however, involve any "backward causation": The connection is logical, not causal. Whether a particular quantum state may legitimately be ascribed to a system at a time is not fixed by its simultaneous dynamical state alone, nor even by the simultaneous dynamical states of all other systems, for the state depends logically on the nature of the interactions it will later undergo. Notice also that, in the present approach, the specification of a complete "experimental arrangement" may be given purely in quantum mechanical terms: It is not necessary to characterize measuring devices or preparation devices as such, and neither is it necessary to describe them and their behavior in classical terms, if this involves going beyond or against a quantum description.

However, the retrospective character of quantum state ascriptions gives rise to a puzzle. On the one hand, the ascription of quantum states is of primary value in making predictions concerning the future behavior of systems. But on the other hand, in the present interpretation, one cannot know the present quantum state of a system until a time when one is already in a position to know whether such a prediction is true or false. How then can the ascription of quantum states be of any value? To resolve this puzzle, notice that a probabilistic prediction made on the basis of a quantum state ascription is of *conditional* form: It specifies the probability of a particular outcome, *given* some outcome or other. For a particular system, this condition may not be satisfied (e.g., a particular particle emerging from the first Stern–Gerlach magnet may not be detected by the second Stern–Gerlach device). Moreover, one cannot know whether the condition is satisfied until the outcome to which the prediction relates has either occurred or failed to occur. But for a large enough collection of similar systems, each subject to the same "preparation" interactions (e.g., an ensemble of particles emerging from the first Stern–Gerlach magnet), it is practically certain that *some* of them will meet the condition of the conditional probabilistic predictions: And this can be known in advance. Hence, although probabilistic predictions made using the quantum state may not be usefully applied to a given system, they can be usefully applied to a large enough collection of similarly "prepared" systems. This, I believe, is the legitimate import of the oft-repeated thesis that a quantum state may be ascribed only to a large ensemble of systems, but not to an individual system.

The preceding considerations illustrate a sense in which the present interpretation treats dynamical states as primary or real, whereas it ascribes only secondary or instrumental significance to quantum states. Since the practice of quantum state ascription then becomes at once so basic to the theory and so restricted in its legitimacy, one may consider how it is at all feasible. What is it about dynamical states and their evolution that makes it possible to ascribe quantum states to systems and, thereby, to correctly infer their likely behavior using the Born rules? What are the special properties of *M*-type interactions that, on the one hand, lead to their naturally being thought of as measurements, and, on the other hand, allow these interactions to play their role in the Born rules? How, finally, is it possible to reconcile the determinate (though indeterministic) outcome of such a measurement interaction with a quantum description of the compound system composed of system and apparatus? Though a fuller answer to these questions must wait until Chapter 3, it is possible to offer a preliminary answer now, in the form of further consideration of measurement, probability, and the Born rules.

The basic characteristic of an *M*-type interaction that allows it to play the role of a measurement is that it establishes a correlation between the initial dynamical state of one system σ (which may consequently be thought of as the **object system**), and the final dynamical state of another system α (which may consequently be thought of as the **apparatus system**). No matter what the initial dynamical state of σ, the final dynamical state of α (if it is influenced by the interaction at all) will be one of a fixed set of **outcome states**, each associated with a distinct value for a dynamical variable $\mathcal{B}$ on α. Moreover, if the initial dynamical state of σ specifies that (what one might call) the measured variable $\mathcal{A}$ on σ has the value a_i, then the final dynamical state of α (if it is altered at all) will specify a corresponding value b_i for $\mathcal{B}$ on α. Both these features accord with what one might call the classical conception of measurement.

In the classical conception, measurement of a dynamical variable of a system requires a physical interaction between that system and a "probe" system. This interaction alters some property of the probe system, and (perhaps after amplification) the change in the probe system manifests itself as a directly perceptible feature of some macroscopic apparatus, such as the position of a pointer on a scale. It is assumed that the measured variable has some precise real value just prior to the measurement, and that the system–probe interaction establishes a correlation between this value and a subsequent "recording" property of an apparatus system. In this way, a classical measurement reveals the value of the measured variable. Since, in the present approach, quantum dynamical variables do not always have precise values, this classical conception must be modified. But it is possible to modify the classical conception without wholly abandoning it. In particular, it is still possible to think of a measurement as involving a normal physical interaction that effects a correlation between the properties of an object system σ and those of an apparatus system α.

The most significant modification required here concerns the (common) case in which the quantity $\mathcal{A}$ has no precise real value on σ prior to its interaction with α. Nevertheless, if the dynamical state of α is altered, then it will come to specify some precise value b_i for $\mathcal{B}$. But this value b_i will not in this case correspond to any prior value a_i of $\mathcal{A}$ on σ – there was none. And this introduces a significant deviation from the classical conception, according to which it is a feature of a well-conducted measurement that a particular result is obtained if *and only if* the measured variable had that value just prior to the measurement. According to the classical conception, if the states of all systems involved are such that the apparatus would certainly record a particular value for the measured variable, then this must be because the object system had that value just before the measurement. But if a properly functioning apparatus can ever record a value which the measured quantity does not possess, then perhaps this need not be so. Perhaps it is possible for a quantity to lack a value even when the states of all systems involved are such that there is some particular value which is certain to be recorded in any measurement of that quantity which succeeds in recording a value. We shall see later (in Chapter 4) that this possibility is indeed realized on the present approach.

This realization involves a further departure from the classical conception of measurement, but the resulting conception still seems to constitute a natural development of the classical conception to accommodate the behavior of quantum systems. At any rate, it is the conception of measurement in quantum mechanics employed by the present approach.

It is now possible to sketch the way in which the present interpretation negotiates the measurement problem: Further details must again wait until Chapter 3. The measurement problem must be faced, since it arises just in case one tries to treat measurement as a quantum mechanical interaction between an object system σ and an apparatus system α, as the present interpretation does. But a closer look reveals that the problem does not now arise in its traditional form, and so does not need to be solved by providing any complicated demonstration of consistency. The key point is to notice that, because of the differential treatment of quantum states and dynamical states, a system may have a dynamical property which is not assigned probability 1 by the Born rules as applied to its quantum state (or to the quantum state of any other system which includes it).

Suppose, for simplicity, that the world is composed of an object system σ and an apparatus system α, and that σ and α interact via a measurement-type interaction. After the interaction, $\sigma \oplus \alpha$ has a dynamical state uniquely determined by its initial dynamical state. Simultaneously, α has one out of a set of accessible final dynamical states. Most such states represent outcomes of the interaction: *Which* of these (if any) is α's actual final dynamical state is not determined by the final dynamical state of $\sigma \oplus \alpha$. None of the final dynamical states accessible to α is wholly classical, since in none of these states does every dynamical variable pertaining to α have a precise real value. But in each outcome state, α does have sufficient dynamical properties for that state to record a definite outcome of the interaction with σ. Thus, if the interaction affects α at all, it produces an indeterministic "selection" of just one of these possible outcomes on α. And this is what one would see if one were to observe α. Such observation may be very indirect, perhaps requiring many further stages of amplification (which may also be described quantum mechanically). But at some point a stage is reached at which a definite recording property of some macroscopic quantum system results which can be directly perceived. That is why quantum measurements are observed to give definite results.

It follows that there is no *conceptual* requirement that an apparatus system should be macroscopic, or be ascribed a definite classical state. But in fact, given human perceptual limitations, any quantum measurement we perform will incorporate some macroscopic system whose different outcome states we can directly distinguish. And though, strictly speaking, no system may correctly be ascribed a definite classical state (with precise values for all dynamical variables), it is at least plausible to suppose that the actual dynamical state of any macroscopic system suited to display the result of a quantum measurement will be observationally indistinguishable from a definite classical state.[1]

What, then, of the projection postulate, or "reduction of the wave packet," of the present approach? In the above account of measurement it was not necessary to suppose that measurements or measurement-type interactions correspond to any distinctive physical process whereby the state of a system changes discontinuously and indeterministically. Dy-

---

[1] The idea here is that the degree to which a quantum dynamical state approximates a classical dynamical state typically increases as the complexity of the underlying quantum system increases. It is thus not macroscopic size but complexity (i.e., number of component quantum systems) which (typically) corresponds to classical behavior. This idea is reminiscent of a supposition entertained by Leggett (1980), that deviations from the normal Schrödinger evolution law may appear and increase with increasing complexity. The chief difference is that on the present interpretation the progressive approach to classical behavior implies no deviation from the quantum dynamical laws, but only deviation from the usual conception of quantum kinematics, which is based on the idea that the dynamical state of a quantum system is given by its quantum state.

namical states do undergo indeterministic as well as deterministic changes, but these indeterministic transitions occur during *any* type of interaction, not just during measurement. Some measurement-type interactions (such as passage through a Stern-Gerlach device) may also serve as preparatory interactions for quantum states, in which case it may be appropriate to ascribe a different quantum state to a system after it has been subjected to such a measurement-type interaction. In that case, the system's quantum state *would* undergo a discontinuous change upon measurement. This would not correspond to any distinctive physical process, however, but simply to a change in perspective. Recall that a system's quantum state does *not* simply represent its dynamical state, but depends on, or is relative to, the interactions it will in fact undergo. The initial quantum state was ascribed to the system relative to its upcoming passage through this measurement/preparation interaction: The subsequent quantum state was ascribed relative to its passage through a *different,* later measurement-type interaction. Note that the perspectives and their change need not be characterized in terms of knowledge, though it is true that this change of perspective *does* correspond to a (potential) change in knowledge: One who comes to know that passage through the measurement/preparation interaction successfully prepared the system to record an outcome in a later measurement-type interaction may represent this change in knowledge via the altered quantum state.

Wave packet reduction is relevant to a further worry concerning measurement. Reduction of the wave packet upon measurement is usually taken to imply that immediate repetition of a measurement of a quantum dynamical variable will certainly (with probability one) yield the same result as before. A difficulty with this implication is that many actual measurements seem to disturb the measured quantum dynamical variable so that immediate repetition may well give a different result. But whether or not this is so, it *is* true that quantum measurements are **verifiable**; that is to say, repeated

43

observation of the "apparatus" system at the conclusion of a measurement interaction reveals it to be recording the same result of the measurement, as long as it is not reset, or otherwise grossly interfered with. This would be a natural consequence of a version of wave packet reduction which "solved" the measurement problem by postulating that a measurement leaves the apparatus system in a quantum state for which the Born rules assign probability one to the recording property corresponding to the value of the object system variable measured. The present interpretation must provide an alternative explanation of the verifiability of measurement outcomes.

The outline of that explanation is as follows. A subsequent observation of the recording property on an apparatus system involves a further measurement-type interaction between that system, now considered as object system, and another apparatus system. One explains the verifiability of measurement outcomes by showing that (and why) this second interaction is certain to give rise to that recording property of the second apparatus system which corresponds to the recording property of the first apparatus system that resulted from the initial measurement-type interaction. It is required of the present interpretation that it be possible to give this demonstration on the basis of assumed laws governing dynamical states: Chapter 3 shows how this requirement is met. Moreover, it also proves possible to distinguish "ideal" measurement interactions, for which an immediate repetition of an initial measurement, by means of a further interaction between an apparatus system and the initial object system, will certainly yield the same result, from "nonideal" interactions, for which it may yield a different result. Interactions of the first type mimic the reduction of the wave packet even more closely, even though neither the quantum state of the object system, nor that of any compound system including it, is such that the Born rules assign probability one to the measured value of the variable measured on the initial object system.

Even accepting that the present interpretation offers a consistent quantum mechanical treatment of measurement in terms of M-type interactions, it remains a puzzle as to why the statistics of outcomes of such interactions should conform to the Born rules. For, in the present interpretation, those statistics are ultimately generated by sequences of indeterministic transitions among dynamical states, taking place throughout the interaction. But the laws governing these supposed transitions, though unknown, are certainly distinct from the Born rules themselves. It turns out, however, that the Born rules follow simply if one imposes a very natural probability metric on the set of dynamical states accessible to a noninteracting subsystem of a compound system with a given dynamical state, at least for certain classes of M-type interactions preceded by P-type interactions. This is shown in Chapter 3. Nevertheless, the mystery is not completely solved, since the underlying indeterministic dynamics are quite unknown, and it remains unexplained why the cumulative effect of multiple indeterministic transitions should result in just this assumed probabilistic connection among the dynamical states of systems and their subsystems. If the present interpretation is correct, then here it is pointing beyond quantum mechanics to a level of unexplored indeterministic processes on which it rests, in a way which is not presently understood.

It may be helpful to say a few words about the interpretation of probability in the present approach. Probability enters on three levels. At the basic level, all interactions induce indeterministic transitions among the dynamical states of the interacting systems. In the (unknown) laws governing these transitions, probability enters as a quantitative measure of a physical disposition – for a system to undergo one (or more) of various possible dynamical transitions during the course of the interaction. Perhaps some frequency analysis of such probabilities will be possible; perhaps a propensity analysis is preferable. The cumulative effect of these indeterministic transitions is to leave each pair of interacting systems in correlated dynamical states, with a definite probability attached to each accessible state. The dynamical state of their compound is then related to the dynamical states of its components by what one might call a **probabilistic law of coexistence**, in accordance with the assumed probability metric.

Probability as it figures in this law of coexistence is understood somewhat differently. Whatever the present dynamical state of the compound system, each of its noninteracting subsystems will in fact presently have some particular dynamical state. The ascription of a probability other than zero or one to several such states is not in conflict with this, but is an instance of the familiar fact that probability statements attribute probabilities to events or states of affairs only under a description. When a coin toss is described as a toss of an evenly balanced coin under normal conditions, there may be probability 1/2 that the coin lands heads up; but for that same coin toss described as a toss of a coin that lands heads up, there is probability 1 that the coin lands heads up. Closer analogs to the present case occur in statistical mechanics. A particular sample of gas may both have a microscopic state lying in a given region of phase space, and simultaneously have probability between zero and one that its microscopic state lies in that region. This latter assertion is not plausibly thought of as expressing a propensity. Its content may be clarified by reference to some hypothetical ensemble of similar samples of gas satisfying the same description as the given sample (for example, with the same energy). The probability corresponds to the (expected) fraction of this ensemble whose microscopic state lies in the given region of phase space. But the content of the probability assertion does not require the *actual* existence of such an ensemble. Similarly, the content of the proposed probabilistic law of coexistence may be clarified by appeal to a purely hypothetical ensemble of similar compound systems, each in the same dynamical state. The probabilities correspond to the (expected) fractions of this ensemble whose components have given dynamical states. The content and meaningfulness of the law of coexistence is unaffected by the fact that in this case there is only one actual compound system.

Finally, the probabilities that appear in the Born rules are conditional probabilities derived from the probabilistic law of coexistence. As such, they do not correspond to transition probabilities – neither from one quantum state to another, nor from an initial classically described apparatus state to a final classically described apparatus state. And if they are taken as quantitative measures of physical dispositions, these cannot be considered primitive physical features: Probability is not plausibly understood to correspond to an unanalyzable propensity. The content of these probabilistic assertions may be clarified by reference to the (expected) fraction of apparatus systems with a particular outcome out of those with any outcome at all, within a hypothetical ensemble of similar compound systems (each including a similar apparatus system) all described by the same dynamical state.

Quantum mechanics is an empirically successful theory which is held to explain a wide variety of otherwise puzzling phenomena. One class of phenomena whose puzzling character has been emphasized by recent experimental as well as theoretical work involves correlations in the results of measurements performed on pairs of systems at a time when the members of each pair are spatially distant and apparently not interacting with each other. Notoriously, certain of these correlations are in violation of a set of inequalities first derived by Bell (1964) as a constraint on any local deterministic hidden variable theory. Further theoretical investigations have both generalized Bell's original inequalities to a class of **Bell inequalities**, and also made more explicit the conditions whose imposition on a theory forces its predictions to conform to these inequalities.

It is an important interpretative task to show how quantum

mechanics explains violations of the Bell inequalities. That task involves more than merely applying the Born rules to the relevant systems to calculate the quantum mechanically predicted correlations, and pointing out that these are inconsistent with the Bell inequalities. It is necessary also to determine which of the conditions that collectively imply the Bell inequalities quantum mechanics fails to satisfy, and to assess the physical and metaphysical significance of this failure. Does quantum mechanics violate any physically or metaphysically defensible locality condition? If not, then what, according to quantum mechanics, gives rise to the observed correlations? More than one influential interpretation of quantum mechanics is unable to satisfactorily answer these questions, as will be shown in Chapter 5. The present interpretation offers a distinctive account of the origins of quantum correlations and of the violation of the Bell inequalities. Chapter 4 gives the technical underpinnings of this account, and Chapter 5 investigates some of its implications for holism and the metaphysics of causation. After a brief description of the puzzling correlations, I shall devote the rest of this overview to a preliminary sketch of how (in the present interpretation) quantum mechanics accounts for them.

Quantum mechanics predicts remarkable correlations in the behavior of systems which may have previously interacted, but are now (apparently) physically and spatially separated. Einstein, Podolsky, and Rosen highlighted some such correlations in a thought-experiment which they presented during their classic (1935) argument against one version of the Copenhagen interpretation. Bohm (1951) described a similar, but theoretically less problematic, thought-experiment which could be used to make the same point. In Bohm's thought-experiment, a pair of spin-½ particles is prepared in a particular quantum state (the singlet state), and then a measurement of a dynamical variable of spin–component along a given axis is made on each of the two particles. In this situation, the Born rules determine probability distributions for all possible results of either a single or a joint spin-component measurement on the two particles. For present purposes, we need note only the following three implications, which follow irrespective of the distance between the particles when the measurements are made. The first applies to measurements on each particle alone:

1. For each particle, and any arbitrary axis, the probability is 1/2 of finding the spin-component of that particle along that axis to be up.

The other two implications concern joint measurements of spin-components of both particles. It is (2) which gives the analog in this situation to the correlations which were crucial to the original EPR argument.

2. As long as the axes are the *same* in each measurement, the measurements will produce a perfect anticorrelation: If one particle is found to have spin up along this (arbitrarily chosen) axis, then the other is certain (probability one) to be found to have spin down, and vice versa.
3. As long as the axes are at an angle of 120 degrees to each other, the probability that both measurements yield the same result (both up, or both down) along their respective axes is 3/4.

In view of the anticorrelation expressed by (2), it is natural to suppose that the two particles have a common source, that each leaves this common source with some property which determines the outcome of a measurement of any spin-component on that particle, and that the properties are themselves anticorrelated for the two particles. (Indeed, the naive realist gives just such an account of the origins of the anticorrelations: For him the property in question is the spin of the particle.) Now the two measurements on each particle pair may occur simultaneously, but arbitrarily far apart (more precisely, they may occur at spacelike separation from one another). And, as Bell (1964) first showed, the natural explanation of the anticorrelations cannot be reconciled with (1) and (3) without conflicting with a principle whose instantiation in the present situation I shall call **Bell locality**.

Bell locality is that the result of the measurement on one particle in the Bohm version of the EPR thought-experiment should not depend on what measurement, if any, is performed on the other particle.

Any probabilities assigned in accordance with Bell locality must satisfy Bell inequalities which are violated by the assignments of (1)–(3).

In a letter to Schrödinger, Einstein discusses the EPR paper, and formulates another principle which may be applied directly to the Bohm version of the EPR thought-experiment, which I shall call **Einstein locality**.

Einstein locality is that "the real state of B... cannot depend on the kind of measurement I carry out on A."[2]

These locality principles are closely related to each other, and each is closely related to the general metaphysical principle that there is no instantaneous action at a distance. In the context of relativistic space-time structure, this principle may be formulated as follows:

NIAD: If two events e_a, e_b occur at spacelike separation from one another, then there can be no direct causal connection between them.

Suppose that e_a is the event which occurs at A when some spin-component of A is selected and measured. And suppose

[2]This and the later passage from Einstein's letter to Schrödinger of June 19, 1935 are quoted in Howard (1985).

50

that e_b is some change in the real (dynamical) state of B. If NIAD holds, then e_a cannot cause e_b if e_a and e_b are spacelike separated. Hence, if NIAD holds, there can be no causal dependence of the real (dynamical) state of B (within a region which is spacelike separated from e_a) on the kind of measurement that is performed on A. Thus explicated, Einstein locality follows from NIAD. Now, insofar as Einstein himself adhered to the naive realist approach to quantum mechanics, he would presumably have believed that the result of a spin-component measurement reveals the real preexisting value of that variable in the Bohm version of the EPR thought-experiment. By this assumption, the result of a measurement of spin-component on B can depend on the kind of measurement performed on A only if the real preexisting value of B's spin-component also depends on the kind of measurement performed on A. In that case, provided that the two events, of selecting and then performing spin-component measurements on A and B, are spacelike separated, Bell locality would follow from Einstein locality. But for an approach which *denies* that measurements simply reveal preexisting values, Bell locality may fail even though Einstein locality holds.

In the sentence immediately preceding that quoted above in his letter to Schrödinger, Einstein formulates another principle which I shall call **Einstein separability**.

Einstein separability: ". . . . the real state of [the two particle compound system] $A \oplus B$ consists precisely of the real state of A and the real state of B, which states have nothing to do with one another."

The last clause of this principle may be intended to exclude causal dependence or descriptive dependence, or both. I shall understand the principle purely as a descriptive principle: So understood, it implies that A and B each have dynamical states of their own, and all the dynamical properties of $A \oplus B$ supervene on dynamical properties of A and of B. (Einstein himself may well not have understood the principle this way, for he takes Einstein locality to follow immediately from Einstein separability, which on my reading it does not, without additional assumptions.) Notice that if this principle were to fail, $A \oplus B$ would have certain irreducible properties, not derivable from the properties of A and B.[3] That is exactly what happens according to the present approach.

Recent experiments by Aspect and others (1982a, 1982b) have verified the quantum mechanical predictions derived from the Born rules in circumstances which are very similar to those of the Bohm version of the EPR thought-experiment. This suggests, but does not prove, that quantum mechanics itself violates some principle of locality, or some principle of separability, or both. A satisfactory interpretation of quantum mechanics should settle this question. If either locality or separability is violated it should provide a satisfactory metaphysical framework for understanding how this violation is possible. And it should exhibit the structure of the explanation quantum mechanics provides of the remarkable correlations manifested in the Aspect-type experiments.

Let e_A (e_B) be an event consisting of the selection and performance of a measurement of spin-component of $A(B)$ along some axis during a joint measurement on both particles in a Bohm pair. Assume first that the events e_A, e_B are not spacelike but timelike separated, and that e_A precedes e_B. Prior to any measurement interactions, A has no definite spin-component in any direction, and neither does B, but in the dynamical state underlying the spin singlet quantum state, the

[3] If one takes the real state of a system to be given by its *quantum* state (as does one popular variant of the Copenhagen interpretation), then quantum mechanics immediately conflicts with Einstein separability, since the singlet quantum state of the two particle system $A \oplus B$ does not consist of the quantum state of A and the quantum state of B. But, of course, Einstein did *not* take the real state of a system to be given by its quantum state. Indeed, one of his strongest reasons for denying that the quantum state of a system gives its real state, and consequently for rejecting this variant of the Copenhagen interpretation, was his desire to interpret quantum mechanics in a way consistent with Einstein separability.

compound system $A \oplus B$ has many irreducible spin properties. In particular, it has many *correlational* properties of a certain form. Each of these correlational properties is mathematically well-defined (each corresponds to its own definite two-dimensional subspace of the four-dimensional tensor product Hilbert space in which the joint quantum state is defined), and each is associated with its own class of measurement-type interactions suitable for its observation. The closest rendering of such a correlational property in English would be as follows: *Either A has spin-down in direction d and B has spin-up in direction d, or A has spin-up in direction d and B has spin-down in direction d.* But this rendering is potentially misleading since it falsely suggests that $A \oplus B$ has such a correlational property only if it has one of its "disjuncts," where each "disjunct" reduces to properties of A and B, separately.

The anticorrelations described in (2) may now be accounted for as follows. (For simplicity, assume that both measurements of spin are nondisturbing.) The measuring interaction with A produces a "recording" property in the measuring apparatus M_A that records either spin-down or spin-up for the spin-component along the axis d' along which M_A is set; it also produces the corresponding spin property in A. The interaction leaves $A \oplus B$ with the correlational property: *Either A has spin-down in direction d' and B has spin-up in direction d', or A has spin-up in direction d' and B has spin-down in direction d'.* It also destroys many other correlational properties. It is important to note that on the present approach the M_A interaction does not change any property of B. In particular, it does not produce in B the spin property (say, *spin-up in direction d'*) anticorrelated to the spin property recorded and produced by M_A. This is important because such alteration in the properties of B resulting from a measurement on A would be in violation of Einstein locality. Moreover, it would involve a direct causal interaction between e_A and an event in which B figures, that occurs no later than e_B. Although this does not conflict with NIAD under the assumption that

53

e_A, e_B are timelike separated, it *would* conflict with NIAD if they were spacelike separated. I take it to be an important advantage of the present approach that its account of violations of the Bell inequalities is consistent with both NIAD and Einstein locality.

The measuring interaction with B is then certain to produce a "recording" property in M_B that records (and produces) spin-up for the spin-component of B along the d' axis just in case M_A recorded (and produced) the result spin-down for A along that axis. The certainty of this outcome (say, spin-up) of M_B is grounded not in any prior property of B itself, but in prior properties of the compound system $A \oplus B$ of which B is a part, namely, the reducible property A *has spin-down in direction* d', *and* B *has the (trivial) spin property that it always has*, together with the irreducible correlational property *Either* A *has spin-down in direction* d' *and* B *has spin-up in direction* d', *or* A *has spin-up in direction* d' *and* B *has spin-down in direction* d'.

Prior to the M_A measurement interaction, the $A \oplus B$ system had the conditional disposition to give result spin-up in an M_B interaction if the M_A interaction gave result spin-down. It was a matter of pure chance that the M_A interaction did give result spin-down: But given that it did, the $A \oplus B$ system then acquired an unconditional disposition to give spin-up if subjected to the corresponding M_B interaction. The conditional disposition is grounded in the correlational property *either* A *has spin-down in direction* d' *and* B *has spin-up in direction* d', *or* A *has spin-up in direction* d' *and* B *has spin-down in direction* d', and the unconditional disposition is grounded in this correlational property together with the further property A *has spin-down in the* d' *direction, and* B *has the (trivial) spin property that it always has*, acquired just in case the M_A interaction gives the result spin-down.

The explanation of the correlations described in (3), and hence of the violations of the Bell inequalities, involves only one more idea: that of a probabilistic disposition. An object has a **probabilistic disposition** to do α in circumstances C just in case, if it were in circumstances C, it would do α with probability p. Ordinary dispositions are special cases of probabilistic dispositions with probability $p = 1$. Prior to the M_A interaction, the $A \oplus B$ system had the conditional probabilistic disposition, with probability 3/4, to give result spin-down (say) in an M_B interaction if subjected to an M_A interaction which gives result spin-down. It was a matter of pure chance that the M_A interaction did give result spin-down: but given that it did, the $A \oplus B$ system then acquired an unconditional probabilistic disposition with probability 3/4 to give spin-down when subjected to the corresponding M_B interaction. These conditional and unconditional probabilistic dispositions are grounded in precisely the same properties of $A \oplus B$ which grounded the dispositions responsible for the anticorrelations described in (2).

It is time now to consider how to extend the above account to cover the case in which e_A and e_B are not timelike but spacelike separated events. For one of the most remarkable features of some Aspect-type experiments is that these events occur at spacelike separation from one another, and, hence, have no invariant time-ordering. Since the correlations themselves are the same whether e_A, e_B are timelike or spacelike separated, it is reasonable to require that their explanation not be radically different in these two cases, even though the above explanation for the timelike case gives temporal and explanatory priority to e_A over e_B. Can an adequate explanation still be given?

I want to suggest the form such an explanation might take. In following this account it will be helpful to refer to the accompanying space-time diagram, Figure 1.1. Consider first how e_A and e_B are causally linked from the perspective of a frame F_1 in which e_A temporally precedes e_B. Let e'_B be the event at B that occurs simultaneously with e_A (according to F_1). Each succeeding moment in F_1 defines an instantaneous spatial location of $A \oplus B$ in F_1. At all these moments (but at

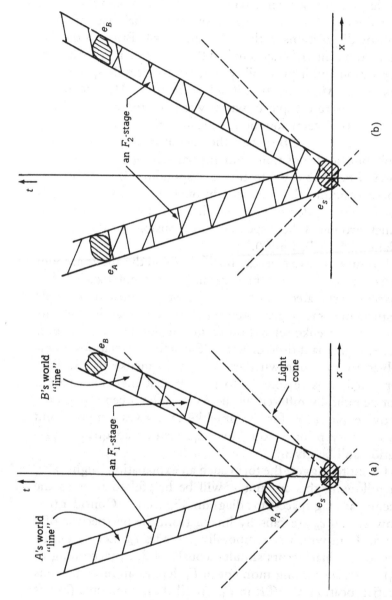

Figure 1.1 Space-time diagrams of Aspect-type experiments: (a) e_A and e_B timelike separated; (b) e_A and e_B spacelike separated.

no preceding F_1-instant) $A \oplus B$ has all those properties that underlie the unconditional disposition of B realized in e_B. Moreover, if e_A had not occurred with the result it gave, $A \oplus B$ would not have had all these properties at F_1-instants succeeding e_A, e'_B. This counterfactual is grounded in a kind of causal relation between e_A and the properties of $A \oplus B$, F_1-subsequent to e_A, e'_B, linking these also to e_B. Together, these causal connections mediate the causal connection between e_A and e_B. A similar account of what mediates the causal connection between e_A and e_B applies for *all* frames F in which e_A occurs earlier than e_B. Moreover, an exactly symmetrical account (with A and B interchanged) of what mediates the causal connection between e_A and e_B applies for every frame like F_2 in which e_B occurs earlier than e_A. All of these accounts are consistent with one another, and all are correct. Together, they provide the full causal explanation of the symmetrical causal connection between e_A and e_B.

This explanation may fail to satisfy, because it seems to mask rather than resolve the problem of how e_A and e_B can be causally connected without violating either locality or NIAD. It is true that the present account (unlike that of the naive realist) explicitly denies that e_A in any way changes the properties of B (and similarly with A and B interchanged): And to that extent, it denies that there are any direct causal connections between spacelike separated events. But even though no dynamical property of B is affected by the selection or outcome at e_A, the subsequent behavior of B at e_B *is* dependent on these: e_B is certain to give a spin–up response just in case e_A occurs with the same axis but a spin–down response. According to the present account there is a causal connection between e_A and e_B which explains this conditional certainty. And does this not already constitute an unacceptable violation of locality and NIAD, whether or not one supposes, as the naive realist does, that e_A affects e_B via an alteration in the properties of B?

The first part of an answer to these questions requires clarification of the relevant locality conditions. The present approach is clearly consistent with Einstein locality. But what about Bell locality? Here it is necessary to further explicate that principle. Bell locality, as originally understood, is a condition which presupposes that the measurement outcome on each particle is strictly determined. But in the present interpretation quantum mechanics is not such a theory. The problem of adequately formulating a condition corresponding to Bell locality for indeterministic theories has been much discussed in recent years. A significant clarification was given by Jarrett (1984). He distinguishes two principles which he calls *locality* and *strong locality,* both of which are applicable to indeterministic theories, and argues persuasively that whereas a valid proof of the Bell inequalities for an indeterministic theory requires strong locality, it is rather locality which is both the natural generalization of Bell locality in the indeterministic case, and also clearly and directly motivated by the requirement that it is impossible to transmit signals between spacelike separated regions. Strong locality is equivalent to the conjunction of locality and another principle Jarrett calls *completeness.* Hence, a failure of an indeterministic theory to satisfy the Bell inequalities may be blamed either on the violation of locality, or on the violation of completeness. As applied to the present circumstances, the principle of locality states, roughly, that the probability of any result of a measurement of B's spin-component is the same, no matter what measurement, if any, is made on particle A. This is true of the quantum mechanical probabilities derived from the Born rules. Completeness states, roughly, that the result of a measurement of a spin-component on B is probabilistically independent of the result of a measurement of spin-component on A. That is clearly not the case for the quantum mechanical probabilities derived from the Born rules. The key question concerning locality is therefore whether or not this failure of completeness *itself* implies any physically or

58

metaphysically worrying action at a distance for the present approach to quantum mechanics.

I shall argue that it does not by defending the physical and metaphysical tenability of an account according to which it is the presence of a certain kind of holistic causal process linking e_A and e_B which explains both this failure of completeness, and the resulting violation of the Bell inequalities. This will also provide the second part of the answer to the questions we were left with earlier. It will be critical to the defense of this account to show that it is indeed consistent with the only justifiable form of NIAD.

Recall that the principle NIAD states that there can be no direct causal connection between spacelike separated events. The qualification "direct" is intended at least to permit pairs of spacelike separated events to be connected through a common cause which is timelike separated from each of them. Now the results of the M_A and M_B interactions are not indirectly connected by any familiar type of common cause, according to the present account. But they are indirectly connected through intermediate causal processes, namely, those states of $A \oplus B$ that in some frame are temporally successive and occur temporally between e_A and e_B. The existence of such processes makes the causal connection between e_A and e_B indirect rather than direct. NIAD may be understood to imply that there can be no *asymmetric* causal connection between spacelike separated events, whereby just one of the events is the cause of an effect at the other. If so, this implication holds for the present account: The causal connection it involves between spacelike separated events e_A and e_B is quite symmetric. The present account conflicts with NIAD only if NIAD is understood to deny the possibility of *any* causal connection between spacelike separated events like e_A and e_B other than their possession of a familiar type of common cause, timelike separated from each. But there is no physical or metaphysical foundation for NIAD as so understood.

59

One might seek a physical foundation for NIAD in relativity theory. Some traditional interpretations of relativity theory have taken that theory to rest on, or at least to imply, NIAD. But more recent analyses of relativity theory[4] have made it clear that the theories of special and general relativity are best formulated as theories of space-time structure, without any reference to causal relations. And when so understood they do not themselves imply NIAD. It might seem that a more promising approach would be to defend NIAD by showing that violations of the principle would lead within relativity theory to causal paradoxes analogous to those which have been argued to plague so-called tachyons – particles with spacelike trajectories. But the construction of such paradoxes relies on the assumption that any causal connection between a pair of spacelike separated events would be *asymmetric,* such that one of the pair is the cause, which may be brought about or independently manipulated to vary the properties of the other (its effect). And this is not true of the events e_A and e_B in the present approach. There is no foundation within relativity theory for any version of NIAD which is in conflict with the present account of violations of the Bell inequalities.

A metaphysical foundation for NIAD may be sought in an analysis of causation. But here again the only justifiable form of NIAD is one which is perfectly consistent with the present account. There are two important metaphysical constraints that might naturally be imposed on any causal relation between noncoincident events, and these together provide the only real metaphysical foundation for NIAD. The first is that there can be no asymmetric causal relation between noncoincident events unless these events are in some correspondingly asymmetric temporal relation. The second is that all pairs of noncoincident causally connected events must be connected by some causal process. The only metaphysically defensible form of NIAD follows from these two principles.

---

[4] See, in particular, Friedman (1983).

On the one hand, since spacelike separated events do not stand in any asymmetric (absolute) temporal relation, it follows that there can be no *asymmetric* causal relation between spacelike separated events. On the other hand, if it were true (as I have argued that it is not) that the only kinds of causal processes capable of linking spacelike separated events link them indirectly through a familiar type of common cause, spacelike separated events could not be causally connected in any other way.

But the present account is consistent with both metaphysical constraints, and so consistent also with the only defensible version of NIAD. There *is* a causal process linking e_A and e_B in the present account (via intermediate states of $A \oplus B$), and so the second constraint is met (in a perhaps unexpected way). And, since in the case in which e_A and e_B are spacelike separated the causal connection between e_A and e_B is taken to be symmetric, there is no conflict with the first constraint either.

One might object that no possible causal process linking e_A and e_B has in fact been described. Certainly, in the case in which they are spacelike separated, e_A and e_B are not linked by (absolutely) temporally successive stages of a process. But that is as it should be: If e_A and e_B are to be connected by any causal process, this cannot consist of a set of (absolutely) temporally successive stages, since e_A and e_B are not asymmetrically temporally related. The present account is based on a new kind of causal process which does *not* simply consist of a set of (absolutely) temporally successive stages. This response is unlikely to convince the objector. Surely, it will be objected, this supposed new kind of causal process would *itself* be in violation of NIAD? For e_A cannot be causally connected to all the happenings that constitute $A \oplus B$'s acquisition and continued possession of irreducible properties at F_1-simultaneities, F_1-subsequent to e_A, since e_A is spacelike separated from at least some of these happenings. But this is where the holism of the account enters in a significant way. $A \oplus B$'s acquisition and continued possession of irreducible properties F_1-subsequent to e_A does not consist of localized happenings coincident with A and with B. To say that these properties of $A \oplus B$ are irreducible is to imply that $A \oplus B$ acquires and possesses them *as a whole*, and irrespective of the acquisition and possession of any properties by A or B, separately. The happenings that constitute $A \oplus B$'s acquisition and continued possession of irreducible properties at F_1-simultaneities F_1-subsequent to e_A cannot be broken down into component happenings, many of which would be spacelike separated from e_A. They are *holistic* happenings, which occupy F_1-successive space-time regions, and none of these regions is, as a whole, spacelike separated from e_A, or from e_B. The new kind of causal process required by the present account is possible only because it consists, in part, of holistic happenings involving irreducible properties of compound systems.

2

Dynamical states

§2.1 THE HIERARCHY OF QUANTUM SYSTEMS

It is frequently noted that some quantum systems are composed of others – their subsystems: For example, a hydrogen atom is composed of a proton and an electron, and an α-particle is composed of two protons and two neutrons. It is less frequently noted that all such quantum systems may also be thought of as subsystems of more extensive systems: Thus, any ten α-particles constitute a quantum system, no matter how widely dispersed – and indeed this system has the important property that its quantum state is required to be symmetric under exchange of any pair of α-particles. In the context of nonrelativistic quantum mechanics, it is reasonable to assume that there is a fixed set of quantum systems, none of whose members has any nontrivial subsystems (i.e., none except itself and the **null subsystem Ø**, which is a subsystem of every quantum system); any system in this set will be called **atomic** (note that atoms are *not* atomic quantum systems in this sense – even a hydrogen atom is composed of an electron and a proton: An electron may be atomic in this sense, but the assumption that it is forms no part of quantum mechanics). All atomic systems together compose the **universal quantum system ω**, which may be finite or infinite, depending on the total number of atomic systems. However, any actual application of quantum mechanics will be to some smaller system which constitutes the universe for the purposes of that application; and it will simplify the subsequent discussion to assume that the universal system ω for any application of quantum mechanics is always composed of some fixed, finite set of atomic systems.

If σ_1 is a (proper) subsystem of σ_2, then we write $\sigma_1 \leq \sigma_2$ ($\sigma_1 < \sigma_2$). The relation $\leq$ partially orders the set Ω of all subsystems of ω. If σ_1, σ_2 are elements of Ω, their least upper bound under $\leq$ is the system $\sigma_1 \oplus \sigma_2$ whose atomic subsystems are the atomic subsystems either of σ_1 or of σ_2. Similarly, their greatest lower bound is the system $\sigma_1 \wedge \sigma_2$, whose atomic subsystems are the atomic subsystems both of σ_1 and of σ_2. Each element σ of Ω has a unique **complement** $\bar{\sigma}$ such that $\sigma \oplus \bar{\sigma} = \omega$ and $\sigma \wedge \bar{\sigma} = \varnothing$. $<\Omega, \leq, \oplus, \wedge, ^- >$ is a finite Boolean algebra: It is isomorphic to the field of all sets of atomic quantum systems. Two systems σ_1, σ_2 are **disjoint** if and only if $\sigma_1 \wedge \sigma_2 = \varnothing$. A set of (nonnull) disjoint systems $\{\sigma_1, \sigma_2, \ldots, \sigma_n\}$ **composes** σ if and only if $\sigma = \sigma_1 \oplus \sigma_2 \oplus \cdots \oplus \sigma_n$. Every subsystem of ω is composed of a unique, finite set of atomic systems, its **atomic components**: The cardinality of this set is the **rank** of the system.

For each quantum system, there is an associated Hilbert space. Dynamical properties of quantum systems correspond to subspaces of their associated Hilbert spaces. When quantum states are introduced (in Chapter 3) it will turn out that a system's quantum state may be represented by a vector, ray, or density operator in its associated Hilbert space. This representation reflects the correspondence between subspaces of a system's Hilbert space and dynamical properties of that system. For example, if a system's quantum state were represented by a vector lying in the unique subspace **P** corresponding to dynamical property $\mathscr{P}$ then the Born rules would assign probability 1 to that dynamical property: In this quantum state, the system is certain to manifest that property (in a way which will become clear in Chapter 3). Though there may be subspaces to which no dynamical property corresponds, every dynamical property corresponds to some subspace, and properties will be individuated in such a way that distinct properties correspond to distinct subspaces. Thus, dynamical properties are closely related to the propositions of the quantum logicians, Mackey's (1963) questions, and Stein's (1972) eventualities. Since for each subspace of a Hilbert space there is a unique projection operator that projects onto that subspace, it is often convenient to focus instead on the one-to-one correspondence between a dynamical property $\mathcal{P}$ and the projection $\mathbf{P}$ onto its corresponding subspace $\mathbf{P}$.

If a quantum system σ is composed of n components $\sigma_1, \sigma_2, \ldots, \sigma_n$, then the Hilbert space H^σ is the n-fold tensor product of the Hilbert spaces of its components: This is written $H^\sigma = H^{\sigma_1} \otimes H^{\sigma_2} \otimes \cdots \otimes H^{\sigma_n}$. If σ is neither null nor atomic, then a key feature of the present interpretation is the division of the set of all possible dynamical properties of σ into two subclasses. A dynamical property of σ is **composite** just in case its corresponding projection operator $\mathbf{P}^\sigma$ is expressible as a direct product $\mathbf{P}^\sigma = \mathbf{P}^{\sigma_1} \otimes \mathbf{P}^{\sigma_2}$ of projections onto the Hilbert spaces $H^{\sigma_1}, H^{\sigma_2}$ of two nontrivial components σ_1, σ_2 of σ ($\sigma = \sigma_1 \oplus \sigma_2$). All other dynamical properties of σ are **prime**. A composite property $\mathcal{P}^\sigma$ of σ is **factorizable into** a set of properties $\{\mathcal{P}^{\sigma_1}, \mathcal{P}^{\sigma_2}, \ldots, \mathcal{P}^{\sigma_n}\}$ ($n \geq 2$) if and only if $\sigma_1, \sigma_2, \ldots \sigma_n$ compose σ and $\mathbf{P}^\sigma = \mathbf{P}^{\sigma_1} \otimes \mathbf{P}^{\sigma_2} \otimes \cdots \otimes \mathbf{P}^{\sigma_n}$. It is easy to show that every dynamical property of a system is either prime or factorizable into some set of prime properties of its components called its **prime factors**. The terminology suggests further that this set is unique. I conjecture that each composite property does indeed possess a unique prime factorization; but the validity of the present interpretation does not hinge on the truth of this conjecture. The conditions governing property ascriptions which will be given in §2.2 ensure that whether or not a system possesses a composite property is always wholly determined by the properties of its components. In this sense, composite properties are always reducible. But prime properties are not generally reducible: Even the entire set of properties of its components does not generally suffice to determine all the prime properties a system has. Thus, as one looks at systems of higher and higher rank in the hierarchy of quantum systems, additional irreducible properties typically appear which these systems do not acquire from the dynamical properties of their components: Such properties may naturally be called emergent.

There are two distinct reasons for distinguishing between prime and composite properties. The first reason is that the distinction is required in order to give a coherent account of measurement interactions: This will be explained in Chapter 3. The second reason is that the consequent division between reducible and irreducible properties is central (in the present interpretation) to an understanding of quantum nonseparability: This is the concern of Chapter 5.

§2.2 THE STRUCTURE OF DYNAMICAL STATES

At each instant, a quantum system has a **dynamical state** which specifies all the quantum dynamical properties it possesses then. The dynamical state may be thought of either as an assignment of truth or falsity to each elementary sentence $\mathcal{A} \in \Delta$, or, alternatively, as specifying, for each dynamical property $\mathcal{P}_A(\Delta)$, whether or not the system has that property. [Here, $\mathcal{A}$ is a quantum mechanical dynamical variable, Δ is a (Borel) set of real numbers, $\mathcal{A} \in \Delta$ is true on σ if and only if the value of $\mathcal{A}$ on σ is restricted to Δ, and $\mathcal{P}_A(\Delta)$ is the dynamical property σ has if and only if the value of $\mathcal{A}$ on σ is restricted to Δ.] Note that this notation presupposes neither that $\mathcal{A} \in \Delta$ is true just in case there is some real number $a \in \Delta$ such that the value of $\mathcal{A}$ is a, nor that $\mathcal{A} \in \Delta$ is true just in case there is some minimal $\Gamma \subseteq \Delta$ such that $\mathcal{A} \in \Gamma$ is true. One may get an initial grasp on this notation by supposing that there is some idealized test which σ would pass if and only if it has $\mathcal{P}_A(\Delta)$, irrespective of whether or not σ would pass a corresponding test for $\mathcal{P}_A(\Gamma)$, for any $\Gamma \subseteq \Delta$. In fact, the properties specified by the dynamical state of any system will always satisfy certain additional consistency conditions which ensure, for example, that if the system has spin up it does not simultaneously have spin down. In addition, there are both statistical and nonstatistical coherence relations among the dynamical states of systems and their subsystems. For example, a system has a composite property whenever it has subsystems which have the properties which are prime factors of that composite property. And, as is shown in Chapter 3, a very natural probabilistic relation between the dynamical state of the universal system and those of its presently non-interacting components underlies the Born rules (at least in certain circumstances). Both the individual consistency conditions and the collective coherence relations governing dynamical states of quantum systems hold because dynamical states satisfy certain basic conditions, which will now be formulated.

The first such condition is called the weakening condition. Quantum dynamical property $\mathcal{Q}$ is said to be **weaker than** property $\mathcal{P}$ just in case $\mathcal{Q}$ corresponds to subspace Q, $\mathcal{P}$ corresponds to subspace P, and Q includes P: This is written $\mathcal{P} < \mathcal{Q}$.

Weakening condition

If σ has property $\mathcal{P}$, then σ also has every property $\mathcal{Q}$ such that $\mathcal{P} < \mathcal{Q}$.

The weakening condition ensures that an important consistency condition on the properties contained in a dynamical state is met. Suppose that $\mathcal{A}$ is a quantum dynamical variable with a range of values V. According to the **property inclusion condition**, if $\triangle \subseteq \Gamma \subseteq V$, then the value of $\mathcal{A}$ is restricted to $\triangle$ only if it is restricted to Γ. For each quantum dynamical variable $\mathcal{A}$ and each $\triangle \subseteq V$, there is a quantum dynamical property $\mathcal{P}_A(\triangle)$ which a system has just in case the value of $\mathcal{A}$ is restricted to $\triangle$. The property inclusion condition holds,

67

then, provided that a system has $\mathcal{P}_A(\Delta)$ only if it has $\mathcal{P}_A(\Gamma)$, whenever $\Delta \subseteq \Gamma \subseteq V$. Quantum dynamical properties correspond to subspaces of Hilbert space, both in conventional interpretations and also for the present interpretation, as subsequent conditions will make clear. It will turn out that the property inclusion condition holds, given the weakening condition.

The properties possessed at any moment by all the systems in the hierarchy may be thought of as being generated by a certain constructive procedure which ensures that the weakening condition is met. The procedure begins by specifying which properties are possessed by atomic systems, and then is expanded inductively to all higher ranks. At each rank, it first specifies which properties a system has by virtue of the properties of its components, and then specifies what further irreducible properties it has. For each system σ, at each instant t, the **system representative** of σ is a subspace $M^\sigma(t)$ of H^σ. If this subspace is one-dimensional, an arbitrary vector which spans it may alternatively be regarded as the system representative. It is important to note that even when the system representative is taken to be a vector, this vector does not necessarily represent the *quantum* state of the system. The present chapter concerns only dynamical states: The notion of a quantum state has yet to be introduced. After the introduction of quantum states in Chapter 3, it will become possible to investigate the circumstances in which the system representative does indeed represent the quantum state of a system. Since all properties of an atomic system are prime, the system representative of an atomic system generates its dynamical state in accordance with the following condition.

System representative condition

If $\mathfrak{R}$ is a prime property of σ corresponding to the subspace R of H^σ, then (provided there is no composite property $\mathfrak{Q}$ of σ which is weaker than $\mathscr{R}$, but which σ does not possess at t) σ has $\mathscr{R}$ at t if the subspace **R** includes $\mathbf{M}^\sigma(t)$.

Indeed, it will turn out that in the presence of the other conditions (and in particular the prime exclusion condition), the system representative condition completely determines the dynamical state of an atomic system. If σ is not an atomic system, then all of its composite properties and perhaps some of its prime properties will be reducible; σ has them by virtue of the properties possessed by its components, in accordance with the following condition, together with the weakening condition.

Composition condition

If σ is composed of $\{\sigma_i\}$ ($1 \leq i \leq n \geq 2$), and $\mathscr{P}^\sigma$ is factorizable into $\{\mathscr{P}^{\sigma_i}\}$, then σ has $\mathscr{P}^\sigma$ if, for all i, σ_i has $\mathscr{P}^{\sigma_i}$.

The composition condition alone determines that σ has certain composite properties: The weakening condition then determines that it has all prime or composite properties weaker than these basic composite properties. Note that since it does not give a necessary condition for possession of a composite property, the composition condition may consistently be imposed even if the conjecture of the previous section is false, so that some composite property possesses two or more distinct prime factorizations. A necessary condition for possession of a composite property is provided by the composite exclusion condition.

Composite exclusion condition

A system has a composite property $\mathscr{Q}$ only if there is some composite property $\mathscr{G} \leq \mathscr{Q}$ which σ has by virtue of the composition condition. The composite exclusion condition per-

69

mits effective application of the proviso in the system representative condition to nonatomic systems. A necessary condition for possession of a prime property is the prime exclusion condition.

Prime exclusion condition

A system has a prime property $\mathcal{R}$ only if *either* it has $\mathcal{R}$ by virtue of the system representative condition, *or* there is some composite property $\mathcal{G} < \mathcal{R}$ which σ has by virtue of the composition condition.

Thus a system inherits all of its composite properties from its subsystems, and acquires many prime properties directly by virtue of its own system representative. But note that a prime property may be ascribed otherwise than through the system representative condition. A system may also inherit further prime properties from its components at one remove, in accordance with the composition and weakening conditions. This happens whenever a prime property is weaker than some composite property which a system possesses by virtue of the properties of its components. (The weakening condition generates no additional properties when applied to prime properties ascribed due to the system representative condition.)

The need for the proviso in the system representative condition arises as follows. It is consistent with all conditions on dynamical states for the system representative to be included in the subspace corresponding to some composite property, which the system does not have because of the composite exclusion condition. Without the proviso, the system representative condition would then lead to inconsistency with the weakening condition.

The application of these conditions may be illustrated in part by Figure 2.1, in which properties are represented by

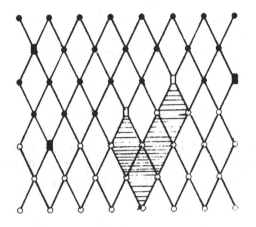

Figure 2.1

nodes, and one property is weaker than another just in case it lies above it along some connected path. The system has a property just in case the corresponding node is filled in. In this diagram, ○ indicates a prime property, □ indicates a composite property, ◇ indicates the property corresponding to the system representative, and shading indicates properties denied by the proviso.

A fuller illustration could be imagined to consist of a vertical stack of trays containing sheets on which there are diagrams similar to this one, where the properties of all systems of a given rank are represented on sheets in the same tray within the stack. Arrows linking a single node in one sheet to several nodes in diagrams in lower trays in the stack would represent the factorizability of the property represented by the former node into the properties represented by the latter nodes.

So far, no restrictions have been laid down governing the relations between the system representative of a compound system and those of its components. But some such restrictions will be needed if properties ascribed directly in accordance with the system representative condition are to mesh

71

correctly with properties ascribed through the composition condition. Now, different degrees of meshing may be required. Perhaps the strongest degree would be that required by the perfect meshing condition.

Perfect meshing condition★

If σ is composed of $\{\sigma^i\}$ ($1 \leq i \leq n$), then the system representative of σ is projected onto by $\mathbf{P}^\sigma = \mathbf{P}^{\sigma_1} \otimes \mathbf{P}^{\sigma_2} \otimes \cdots \otimes \mathbf{P}^{\sigma_n}$, where $\mathbf{P}^{\sigma_i}$ projects onto the system representative of σ_i.

In fact, this condition is *too* strong: To impose it would be to abandon the idea that compound systems may have irreducible properties not dictated by the properties of their components. Thus, this condition will not be imposed within the present interpretation, as is indicated by an asterisk. The following weaker condition is imposed instead.

Minimal meshing condition

If σ is composed of $\{\sigma_i\}$ and also of $\{\tau_j\}$ ($1 \leq i \leq m$, $1 \leq j \leq n$), and the system representatives of σ_i, τ_j are projected onto by $\mathbf{P}^{\sigma_i}, \mathbf{P}^{\tau_j}$, respectively, then $(\mathbf{P}^{\sigma_1} \otimes \mathbf{P}^{\sigma_2} \otimes \cdots \otimes \mathbf{P}^{\sigma_m}) \cdot (\mathbf{P}^{\tau_1} \otimes \mathbf{P}^{\tau_2} \otimes \cdots \otimes \mathbf{P}^{\tau_n}) \neq 0$. In particular, $\mathbf{P}^\sigma \cdot (\mathbf{P}^{\sigma_1} \otimes \mathbf{P}^{\sigma_2} \otimes \cdots \otimes \mathbf{P}^{\sigma_m}) \neq 0$.

The minimal meshing condition ensures that a system is never ascribed incompatible properties. To explain this, and to examine the structure of the set of properties of a quantum system ascribed in accordance with the conditions imposed so far, it is necessary to introduce some further notions. For each dynamical property $\mathscr{P}$ corresponding to projection operator $\mathbf{P}$, there exists another dynamical property $\overline{\mathscr{P}}$, the **negation** of $\mathscr{P}$, corresponding to the projection operator $\mathbf{I}-\mathbf{P}$.

Such a property always exists, since if $\mathcal{P}$ is a dynamical property there must be some dynamical variable $\mathcal{A}$ and set of values $\triangle$, such that σ has $\mathcal{P}$ if and only if the value of $\mathcal{A}$ on σ is restricted to $\triangle$: Then $\overline{\mathcal{P}}$ is the property σ has if and only if the value of $\mathcal{A}$ is restricted to $\mathbb{R}-\triangle$. It is easy to see that according to the above conditions it is possible for a system to have neither a property not its negation. Consider an atomic system α with system representative $\mathbf{M}^\alpha$, and an irreducible property $\mathcal{P}$ such that $\sim (\mathbf{P}>\mathbf{M}^\alpha)$ and $\sim (\mathbf{P}^\perp>\mathbf{M}^\alpha)$, where we write $\mathbf{A}>\mathbf{B}$ just in case the subspace $\mathbf{A}$ properly includes the subspace $\mathbf{B}$. In such a case, an associated dynamical variable $\mathcal{A}$ does not have a precise real value.

Two different notions of compatibility may be distinguished. A pair of properties $\mathcal{P},\mathcal{Q}$ will be said to be **cogenerable** if and only if the commutator of their corresponding projections vanishes: $[\mathbf{P},\mathbf{Q}] = \mathbf{0}$. A set of properties is cogenerable just in case it is pairwise cogenerable. The term is chosen because (when the proviso is met) the system representative condition can directly ensure that each property $\mathcal{P}$ in a cogenerable set is **determinate** – that is, the system possesses either $\mathcal{P}$ or $\overline{\mathcal{P}}$. Cogenerable properties correspond to subspaces which generate a Boolean sublattice of the (non-Boolean) lattice of subspaces of the system's Hilbert space under the subspace operations of intersection, span, and orthocomplementation; and all cogenerable properties become determinate when the system representative is (or is a subspace of) an atom in this Boolean lattice.

Once the perfect meshing condition fails, one must expect that a system can come to possess a noncogenerable set of properties. This possibility underlines the inappropriateness of the term 'incompatible' for noncogenerable properties in the present interpretation. But provided that the minimal meshing condition holds, a system never possesses a pair of properties $\mathcal{P},\mathcal{Q}$ such that $\mathbf{P}\cdot\mathbf{Q}=\mathbf{0}$. In particular, a system never possesses both a property and its negation. A pair of properties $\mathcal{P},\mathcal{Q}$ such that $\mathbf{P}\cdot\mathbf{Q}\neq\mathbf{0}$ will be called **cotenable**. A set of properties ascribed in accordance with the above conditions (including minimal meshing, but not perfect meshing) will always be cotenable, but will not always be cogenerable: This is shown in the appendix.

The set of dynamical properties of a system ascribed in accordance with the preceding conditions has an interestingly nonclassical structure, as may be seen by examining the set of all properties of the form $\mathscr{P}^A(\Delta)$, for fixed dynamical variable $\mathscr{A}$ and arbitrary Borel set Δ. Classically, $\mathscr{A}$ is supposed to have some precise real value r on any system σ to which it pertains. Consequently, for any set Δ, σ has $\mathscr{P}^A(\Delta)$ if and only if $r \in \Delta$ on σ. The following individual consistency conditions on properties of σ all follow from this condition.

Precise value. $\exists r \in \mathbb{R} : \mathscr{A} \in \{r\}$
Set value. $\exists \Delta \neq \varnothing \subseteq \mathbb{R} : \mathscr{A} \in \Delta$
Property inclusion. If $\mathscr{A} \in \Gamma$, and $\Gamma \subseteq \Delta$, then $\mathscr{A} \in \Delta$
Property compatibility. If $\mathscr{A} \in \Gamma$, then $\mathscr{A} \notin \mathbb{R}\text{-}\Gamma$
Property intersection. If $\mathscr{A} \in \Gamma$ and $\mathscr{A} \in \Delta$, then $\mathscr{A} \in \Gamma \cap \Delta$
Dichotomy. $\mathscr{A} \in \Delta$ or $\mathscr{A} \in \mathbb{R} - \Delta$
Minimal set value. $\exists \Gamma \neq \varnothing \subseteq \mathbb{R}: \mathscr{A} \in \Gamma$ and if $\mathscr{A} \in \Delta$ then $\Gamma \subseteq \Delta$.

Set value, property inclusion, and property compatibility still hold for properties ascribed in accordance with the above conditions, but all the other listed conditions sometimes fail according to the present interpretation. The failure of precise value and dichotomy is not unexpected, although it does raise the problem of explaining why properly conducted measurements always seem to yield precise real values as results. That problem will be treated and resolved in the light of the treatment of measurement to be given in Chapter 3. But the failure of the other two conditions raises a further interesting problem which must be faced immediately.

It is natural to suppose that even if the value of a quantity is not a real number, then it is at least some *set* of real numbers – the minimal set to which its value is restricted.

74

This supposition is expressed not by set value, but by minimal set value. But if minimal set value holds, then, given property inclusion, the value of $\mathscr{A}$ is restricted to Δ if and only if $\Gamma \subseteq \Delta$. It follows that if $\mathscr{A} \in \Delta_1$ and $\mathscr{A} \in \Delta_2$, then $\Gamma \subseteq \Delta_1 \cap \Delta_2$, and so $\mathscr{A} \in \Delta_1 \cap \Delta_2$. But this is just the property intersection condition, which does not generally hold. Consider a situation like that illustrated in Figure 1.1 and suppose $\mathscr{P}^A(\Delta_1 \cap \Delta_2)$ corresponds to the ' $\Diamond$,' and $\mathscr{P}^A(\Delta_1)$, $\mathscr{P}^A(\Delta_2)$ correspond to shaded nodes lying above it along some connected paths. By countenancing the possibility of the failure of property intersection, the present interpretation must therefore accept that even when the value of a variable is restricted to a set, which might have been thought of as the *set-value* of the variable, there need be no minimal set to which its value is restricted. But what is the content of the claim that σ has $\mathscr{P}^A(\Delta_1)$ if no value (not even a set) may be ascribed to $\mathscr{A}$ on σ?

The content of the claim is best unpacked by examining the inferences which may be drawn from it. One such inference is that if one were to observe whether or not σ has $\mathscr{P}^A(\Delta_1)$, one would find that it does: And if one were to conduct a less precise observation as to whether or not σ has $\mathscr{P}^A(\Delta')$ (with $\Delta_1 \subset \Delta'$), one would find that it does. On the other hand, there is no *maximally* precise observation of $\mathscr{A}$ which would locate its value with maximal precision (such that any more precise observation would yield a negative result). And perhaps more surprisingly, even though in these circumstances simultaneous observations as to whether or not σ has $\mathscr{P}^A(\Delta_1)$ and $\mathscr{P}^A(\Delta_2)$ are certain to give positive results, one cannot infer that another (more precise) simultaneous observation of $\mathscr{P}^A(\Delta_1 \cap \Delta_2)$ would also give a positive result. It follows that here one cannot reliably observe whether σ has $\mathscr{P}^A(\Delta_1 \cap \Delta_2)$ just by simultaneously observing whether σ has $\mathscr{P}^A(\Delta_1)$ and whether σ has $\mathscr{P}^A(\Delta_2)$. But this is an assumption which one might quite unreflectively make, and so it is necessary to understand the extent of its reliability. There seems to be a twofold reason for this assumption's proving as reliable as it is. For most macroscopic systems, observations of dynamical variables are already very imprecise. Hence, properties like $\mathscr{P}^A(\Delta_1)$ and $\mathscr{P}^A(\Delta_2)$ will be represented by nodes far removed from that corresponding to the system representative of the macroscopic system. Consequently, $\mathscr{P}^A(\Delta_1 \cap \Delta_2)$ are practically certain to be possessed by σ provided that $\mathscr{P}^A(\Delta_1)$ and $\mathscr{P}^A(\Delta_2)$ are. For microscopic systems, on the other hand, observation does not in any case typically reveal the property the system previously possessed, and indeed it is not typical for a system to possess interesting properties of the forms $\mathscr{P}^A(\Delta_1)$, $\mathscr{P}^A(\Delta_2)$ prior to their observation. Consequently, one does not generally infer from positive results of observation to prior possession of $\mathscr{P}^A(\Delta_1)$ and $\mathscr{P}^A(\Delta_2)$, still less to prior possession of $\mathscr{P}^A(\Delta_1 \cap \Delta_2)$. And the situation in which one *would* infer from results of observation to prior possession of $\mathscr{P}^A(\Delta_1)$ and $\mathscr{P}^A(\Delta_2)$ would typically be that in which observations of $\mathscr{P}^A(\Delta_1)$ and $\mathscr{P}^A(\Delta_2)$ are known to be "reliable," just because observation of $\mathscr{P}^A(\Delta_1 \cap \Delta_2)$ is known to be "reliable" – and then the inference would indeed be reliable, if only one needed to make it!

Having said this much in defense of an interpretation that violates property intersection, let me admit to acute discomfort at the need to violate it. This is perhaps the clearest sign that the present conditions on property ascriptions are not yet completely correct. If not, the program would be to modify these conditions to avoid this violation while still preserving the advantages of the present conditions, as these emerge, especially in Chapters 3 and 5.

So far, the notion of the system representative has been very abstract; and the only relations between the representatives of systems of different ranks in the hierarchy of quantum systems have come from the restrictions imposed by the meshing conditions. The content and interrelations of system representatives may be further specified in the following way.

In the present interpretation, it is postulated that the system representative of the universal quantum system ω is one-dimensional, and is at all times spanned by a vector ψ^ω in the appropriate tensor product Hilbert space H^ω. It is then a consequence of the free evolution postulate, to be introduced in the next section, that this evolves continuously and deterministically. All other (nonnull) quantum systems are subsystems of the universal quantum system, and are characterized by their system representatives (which may or may not be one-dimensional). Such systems interact with one another from time to time; their interaction is marked by the presence of interaction terms in the universal Hamiltonian. External interactions induce indeterministic transitions in the representatives of systems subject to them. If a system is subject to no external interactions, its representative is always selected from a privileged set of accessible subspaces of its Hilbert space in accordance with what will be called the subspace decomposition condition. In order to present this condition, it is useful first to state the following interesting lemma, given by von Neumann (1932) (but credited by him to Schmidt), which lies at the heart of Kochen's recent approach to the interpretation of quantum mechanics.

Biorthogonal decomposition lemma

Any vector $\psi^{\alpha\beta}$ in the tensor product Hilbert space $H^{\alpha\beta} = H^\alpha \otimes H^\beta$ may be decomposed as follows

$$\psi^{\alpha\beta} = \sum_i a_i(\chi_i^\alpha \otimes \chi_i^\beta),$$

where the a_i are nonzero complex coeffficients, $(\chi_i^\alpha, \chi_j^\alpha) = (\chi_i^\beta, \chi_j^\beta) = \delta_{ij}$ (which equals 1 if $i = j$, and zero otherwise), and the sum is over an index set I of cardinality not greater than min $\{\dim(H^\alpha), \dim(H^\beta)\}$ [$\dim(H)$ is the dimension of Hilbert space H]. Moreover, this decomposition is unique (up to phase and order of summation) if and only if $|a_i|^2 \neq |a_j|^2$ for $i \neq j$. A coefficient a_i is said to be **degenerate** if and only if $|a_i|^2 = |a_j|^2$ for $i \neq j$. I may be partitioned into subsets I_k such

77

that for all $i, j \in I_k$, $|a_i|^2 = |a_j|^2$: Unit sets I_k correspond to non-degenerate coefficients. For each set I_k there is a unique pair of correlated subspaces $M_k^\alpha \leq H^\alpha$, $M_k^\beta \leq H^\beta$, such that for $i \in I_k$, *either* the vectors χ_i^α may be chosen from an arbitrary orthonormal basis for M_k^α, *or* the vectors χ_i^β may be chosen from an arbitrary orthonormal basis for M_k^β, but not both. The set of all subspaces $\{M_k^\alpha\}$ ($\{M_k^\beta\}$) may be called the **privileged set** of subspaces for H^α (H^β) corresponding to $\psi^{\alpha\beta}$. Any two subspaces in the privileged set for H^α (H^β) are orthogonal.

The subspace decomposition condition may now be stated as follows.

Subspace decomposition condition

Let σ be a quantum system which is not interacting at time t, since the universal Hamiltonian H^ω is of the form $H^\omega = (H^\sigma \otimes I^{\bar\sigma}) + (I^\sigma \otimes H^{\bar\sigma})$. Then the system representative $M^\sigma(t)$ of σ at t is one of the subspaces M_k^σ from the privileged set for H^σ corresponding to the biorthogonal decomposition of any vector ψ^ω spanning the system representative $M^\omega(t)$ onto the Hilbert spaces H^σ, $H^{\bar\sigma}$; and the system representative $M^{\bar\sigma}(t)$ of $\bar\sigma$ at t is the correlated subspace $M_k^{\bar\sigma}$ from the privileged set for $H^{\bar\sigma}$: Moreover, the probability that σ has system representative M_k^σ at t equals $\Sigma_{i \in I_k} |a_i|^2$, where the a_i are coefficients appearing in some biorthogonal decomposition of ψ^ω onto H^σ, $H^{\bar\sigma}$. A subspace in the privileged set for H^σ corresponding to the biorthogonal decomposition of any such vector ψ^ω onto H^σ, $H^{\bar\sigma}$ is said to be **accessible for** σ at t.

It is easy to see that the subspace decomposition condition conflicts with the perfect meshing condition. Thus, consider a simple model universe $\omega = \alpha \oplus \beta$ composed of two atomic systems α, β at a time when they are not interacting. If the system representative of ω is spanned by a vector ψ^ω such that the decomposition

$$\psi^\omega = \sum_{i \in I} a_i(\chi_i^\alpha \otimes \chi_i^\beta)$$

is essentially unique, then the system representatives of ω, α, β are all one-dimensional, and projected onto by $\mathbf{P}_{\psi^\omega}$, $\mathbf{P}_{\chi_i^\alpha}$, $\mathbf{P}_{\chi_i^\beta}$, respectively (for some $i \in I$). But, clearly, $\mathbf{P}_{\psi^\omega} \neq \mathbf{P}_{\chi_i^\alpha} \otimes \mathbf{P}_{\chi_i^\beta}$ unless $|a_i|^2 = 1$: So, the perfect meshing condition fails in any nontrivially superposed state ψ^ω. It is less obvious, but nevertheless true, that there is *no* conflict between the subspace decomposition condition and the minimal meshing condition, as is shown in the appendix.

§2.3 DYNAMICS

In the present interpretation, how do the dynamical properties of a quantum system change as time passes? Consider first the universal quantum system ω. The system representative of ω is a one-dimensional subspace of $\mathbf{H}^\omega$ spanned by a vector ψ^ω. Since ω is the universal system, it is supposed to never interact with any external system, and the system representative of ω is therefore postulated to evolve continuously and deterministically at all times, in a way which would correspond to a unitary Schrödinger evolution of the (Schrödinger picture) vector ψ^ω. (But note that ψ^ω cannot be the *quantum* state vector of ω, since quantum states are legitimately ascribable only to systems about to undergo an appropriate type of external interaction, whereas ω is supposed not to be subject to any external interactions.) This evolution of the system representative of ω determines how the irreducible properties of ω vary with time: Variation of the reducible properties of ω depends on how the representatives of subsystems of ω evolve.

If a system σ<ω is free of interactions throughout an interval (as marked by the absence of interaction terms in the universal Hamiltonian coupling it to other systems), then its system representative is a subspace from the privileged set corresponding to the decomposition of ψ^ω at each instant in that interval. Moreover, the evolution of the privileged set is determined by the deterministic evolution of ψ^ω. This still leaves open whether or not the evolution of the individual subspace of σ is *itself* deterministic. It is therefore necessary to postulate that this is so.

Free evolution

The representative of a system σ evolves continuously and deterministically throughout any interval during which σ does not interact with other systems. More specifically, there exists a unitary operator $\mathbf{U}^\sigma$ $(t_1 - t_0)$ such that if the system representative of σ at t_0 is spanned by the vectors $\{\chi_i^\sigma(t_0)\}$ $(i \in I_k)$, then the system representative of σ at t_1 is spanned by the vectors $\{\mathbf{U}^\sigma(t_1 - t_0) \cdot \chi_i^\sigma(t_0)\}$.

Now, there might seem to be two ways in which this postulate could lead to inconsistency with the subspace decomposition condition. In general, the dimensions of the subspaces in the privileged set of subspaces for $\mathbf{H}^\sigma$ will vary with time. But suppose that $\mathbf{H}^\sigma$ is finite dimensional, and all subspaces accessible for σ at t_0 are two-dimensional, whereas all subspaces accessible for σ at $t_1 > t_0$ are one-dimensional: Then it would be inconsistent with the subspace decomposition condition to postulate that the system representative of σ evolves deterministically from t_0 to t_1. And even if the dimensions of the subspaces accessible for σ do not vary with time, if the probability corresponding to a continuously evolving accessible subspace were to vary, this would imply that the system representative could make transitions into or out of that subspace. But it is easy to show that no inconsistency in fact arises in either of these two ways. Suppose that at the beginning of an interval (t_0, t_1) the system representative of ω is (spanned by) a vector with the following biorthogonal decomposition

$$\psi^{\omega}(t_0) = \sum_{i \in I} a_i(t_0) \cdot (\chi_i^{\sigma}(t_0) \otimes \chi_i^{\bar{\sigma}}(t_0)).$$

Then the probability of the system representative for σ spanned by the vectors $\{\chi_i^{\sigma}(t_0)\}$ $(i \in I_k)$ is given by $\sum_{i \in I_k} |a_i(t_0)|^2$ (for a nondegenerate coefficient, I_k is a unit set). If σ does not interact with any subsystem of $\bar{\sigma}$ during this interval, the Hamiltonian for ω may be written as $\mathbf{H}^{\omega} = \mathbf{H}^{\sigma} \otimes \mathbf{I}^{\bar{\sigma}} + \mathbf{I}^{\sigma} \otimes \mathbf{H}^{\bar{\sigma}}$. Hence, the evolution of $\psi^{\omega}(t)$ throughout this interval will be generated by a unitary operator $\mathbf{U}^{\omega}(t) = \mathbf{U}^{\sigma}(t) \otimes \mathbf{U}^{\bar{\sigma}}(t)$ which is the tensor product of independent evolution operators for σ and $\bar{\sigma}$, respectively. Hence, we have

$$\psi^{\omega}(t_1) = \mathbf{U}^{\sigma}(t_1 - t_0) \cdot \psi^{\omega}(t_0) =$$

$$\sum_{i \in I} a_i(t_0) \cdot [\mathbf{U}^{\sigma}(t_1 - t_0)\chi_i^{\sigma}(t_0) \otimes \mathbf{U}^{\bar{\sigma}}(t_1 - t_0)\chi_i^{\bar{\sigma}}(t_0)].$$

But this already expresses $\psi^{\omega}(t_1)$ as a decomposition of the form

$$\psi^{\omega}(t_1) = \sum_{i \in I} a_i'(t_1) \cdot (\chi_i'^{\sigma}(t_1) \otimes \chi_i'^{\bar{\sigma}}(t_1)),$$

with $\chi_i'^{\sigma}(t_1) = \mathbf{U}^{\sigma}(t_1 - t_0) \cdot \chi_i^{\sigma}(t_0)$, $\chi_i'^{\bar{\sigma}}(t_1) = \mathbf{U}^{\bar{\sigma}}(t_1 - t_0) \cdot \chi_i^{\bar{\sigma}}(t_0)$, and $a_i'(t_1) = a_i(t_0)$. This decomposition is biorthogonal, since $\mathbf{U}^{\sigma}, \mathbf{U}^{\bar{\sigma}}$ are unitary operators: Moreover, the kth subspace in the privileged set for $\mathbf{H}^{\sigma}$ ($\mathbf{H}^{\bar{\sigma}}$) at t_1 has the same dimension as the kth subspace in the privileged set for $\mathbf{H}^{\sigma}$ ($\mathbf{H}^{\bar{\sigma}}$) at t_0. By the subspace decomposition condition, the probability of the system representative spanned by vectors $\{\chi_i'^{\sigma}(t_1)\}$ $(i \in I_k)$ is given by $\Sigma_{i \in I_k} |a_i'(t_1)|^2 = \Sigma_{i \in I_k} |a_i(t_0)|^2$. And so it is consistent with the subspace decomposition condition to require that, provided σ is subject to no external interactions in the time

81

interval (t_0, t_1), if the system representative of σ at t_0 is spanned by vectors $\{\chi_i^\sigma(t_0)\}$ $(i \in I_k)$, then the system representative of σ at t_1 is spanned by vectors $\{\chi_i'^\sigma(t_1)\}$ $(i \in I_k)$, where $\chi_i'^\sigma(t_1) = \mathbf{U}^\sigma(t_1 - t_0) \cdot \chi_i^\sigma(t_0)$. And this is precisely what the free evolution postulate claims.

There is at least one further stability condition constraining the evolution of the representative of a system, which may in certain circumstances suffice to render this evolution deterministic. This condition will prove important in the account of measurement interactions for the present interpretation.

Stability condition

Suppose that σ is subject to an external interaction which transforms the system representative of ω as follows

$$\sum_{i \in I} a_i \cdot (\chi_i^\sigma \otimes \chi_i^{\bar{\sigma}}) \to \sum_{i \in I} a_i \cdot (\chi_i'^\sigma \otimes \chi_i'^{\bar{\sigma}}),$$

where each expansion is biorthogonal, and there is some $\mathcal{A}^\sigma$ such that, for all $i \in I$, $\mathbf{A}^\sigma \chi_i^\sigma = a_i \chi_i^\sigma$, $\mathbf{A}^\sigma \chi_i'^\sigma = a_i \chi_i'^\sigma$. Then, if it is true for the system representative of σ before the interaction that $\mathcal{A} = a_i$, it is also true for the system representative of σ after the interaction.

In general, however, if a system is subject to external interactions, then its system representative may change indeterministically. The present interpretation of quantum mechanics is silent on the details of such indeterministic transitions, except to assume that their probabilities vary with the strength of the interaction, becoming negligible as the interaction strength approaches zero. In the present interpretation, only the cumulative effect of these indeterministic transitions is specified by quantum mechanics; and this is to leave a system at the conclusion of an interaction with a system representative from the privileged set, whose probability is given by the subspace decomposition condition. Moreover, in the present interpretation, these probabilities for being left with a particular system representative at the conclusion of an interaction are the ultimate source of *all* quantum mechanical probabilities. This may seem surprising, since quantum mechanics is usually postulated to yield probabilities of measured results on a system in a given quantum state, and neither measurement nor quantum states have yet been mentioned. It will still hold for the present interpretation that quantum mechanics yields probabilities of measurement results on systems in a given quantum state. However, this is not secured by postulation, but must instead be derived through an analysis of the nature of a measurement interaction and the conditions under which quantum states may be assigned to systems. Such analysis occupies the next chapter: Only at its conclusion will the traditional quantum probabilistic algorithm emerge.

3

Measurement and quantum states

§3.1   MEASUREMENT INTERACTIONS

In the present interpretation terms such as 'measurement,' 'observation,' 'find,' 'classical,' and 'apparatus' do not appear in the formulation of quantum mechanics. Yet it must of course be true that quantum mechanics can be used to predict features of the results of measurements – in particular their statistics. What is required, then, is an account in quantum mechanical terms of those physical interactions capable of being used to perform measurements of quantum dynamical variables, from which it follows that the predicted probabilities for specified behavior of certain quantum systems in these interactions coincide with what, on a more conventional interpretation, are simply *postulated* as the quantum mechanical probabilities for various measurement results on a system with a given quantum state. It is necessary to see how this consequence may be derived, not only in order to explicate the Born rules, but also to explain the role of quantum states within the present interpretation. The first step toward the required account is to analyze the nature of an interaction capable of being used to perform a measurement of the value of a quantum dynamical variable in a quantum system. Any such interaction will be called an **M-type** interaction, whether or not it is in fact used to perform a measurement.

The basic characteristic of an *M*-type interaction is that it establishes a correlation between the initial dynamical state of one quantum system σ (the object system) and the final dynamical state of another quantum system α (the apparatus system). Though σ and α will usually be distinct systems, in some cases an M-type interaction may correlate initial and final dynamical states of a single system, for example, by coupling its prior spin to its subsequent position. What kind of interaction is capable of establishing the needed correlation?

One could view this question as a request for a single, precise definition of the class of M-type interactions. But though I shall offer definitions of certain idealized kinds of M-type interaction, there is no reason to suppose that these exhaust the entire class of M-type interactions. In the present interpretation, that class may be fairly heterogeneous, its various subclasses having only two central features in common: the ability to set up a suitable correlation between dynamical states, and the fact that interactions in all these subclasses support the ascription of a quantum state to one interacting system and its application via the Born rules. Although it may be of some interest to conduct a systematic theoretical investigation of the entire class of M-type interactions, I shall not do so in this monograph. What I shall do is to show that there can be at least *some* M-type interactions, by precisely characterizing one simple variety of M-type interaction and then showing how interactions of this type would indeed establish the right kind of correlation between dynamical states, and would also support the ascription of a quantum state and its use in the Born rules.

What I shall call a simple M-type interaction involving a system σ proceeds in accordance with a characteristic interaction Hamiltonian linking σ to a suitable M-type system α (which may consequently be thought of as an apparatus for the measurement). To say when a simple M-type interaction involving σ occurs, it is therefore necessary to specify the nature of this interaction Hamiltonian, the general features of an M-type system, and the required condition of the particular M-type system α at the start of its interaction with σ.

An interaction between two quantum systems σ and α is **M-suitable for** $\mathscr{A}$ just in case it is characterized by an interaction term $\mathbf{H}_I$ in the universal Hamiltonian which is zero outside an interval (t_0, t_1), and would produce the following Schrödinger evolution in a (hypothetical) compound quantum state vector of $\sigma \oplus \alpha$ over the interval (t_0, t_1), for each element in some orthonormal basis $\{\chi_{ij}^\sigma\}$ of vectors for H^σ with $\mathbf{A}\chi_{ij}^\sigma = a_i \chi_{ij}^\sigma$, and some subspace P_0^α of vectors in H^α,

$$\chi_{ij}^\sigma \otimes \chi_0^\alpha \rightarrow \xi_{ij}^\sigma \otimes \chi_{ij}^\alpha, \qquad (3.1)$$

where $(\xi_{ij}^\sigma, \xi_{i'j'}^\sigma) = (\chi_{ij}^\alpha, \chi_{i'j'}^\alpha) = \delta_{ii'} \cdot \delta_{jj'}$, $\chi_0^\alpha \in \mathsf{P}_0^\alpha$. Note that it is *not* assumed that σ may be ascribed some quantum state vector (either χ_{ij}^σ or anything else) at t_0: Equation (3.1) is simply a (rather weak) mathematical restriction on the form of $\mathbf{H}_I$.

For an M-type interaction to actually occur, it is not enough that a pair of systems interact via an M-suitable interaction: One of these systems must be an M-type system which is M-ready at the start of the interaction. A system is **M-ready for $\mathscr{A}$** at t if and only if its system representative at t is a vector χ_0^α from the subspace P_0^α for which the interaction is M-suitable for $\mathscr{A}$. Suppose that an interaction occurs between σ and α which is M-suitable for $\mathscr{A}$. Let $\mathsf{P}_i^\alpha \leq \mathsf{H}^\alpha$ be defined as $[\chi_{ij}^\alpha : i \text{ fixed}]$: $\mathscr{P}_i^\alpha$ is the **recording property** of α corresponding to the value a_i of $\mathscr{A}$ on σ. (Here and elsewhere the notation [   ] indicates the subspace spanned by the vectors in brackets.) The value a_i of $\mathscr{A}$ is manifested in an M-suitable interaction just in case α has the recording property $\mathscr{P}_i^\alpha$ at the conclusion of the interaction.

Then α is an M-type system if and only if

i.   the $\mathscr{P}_i^\alpha$ are all prime properties of α, and
ii. there is no nontrivial composite property $\mathfrak{Q}^\alpha$ of α such that, for some i, $\mathscr{P}_i^\alpha \leq \mathfrak{Q}^\alpha$.

[Note that (i) is actually redundant, since it follows from (ii): I am indebted to Howard Stein for pointing this out to me.]

Finally, a **simple M-type interaction (SMI) for $\mathscr{A}$** on σ occurs between t_0 and t_1 if and only if σ and α interact through an interaction which is M-suitable for $\mathcal{A}$, where α is an M-type system which is M-ready for $\mathcal{A}$ at t_0.

It follows that distinct recording properties $\mathcal{P}_i^\alpha$, $\mathcal{P}_j^\alpha$ are non-cotenable. Hence, any set of distinct real numbers $\{q_i\}$ defines the possible values of a recording quantity $\mathfrak{Q}$ which corresponds to the self-adjoint operator $\mathbf{Q}^\alpha = \sum_i q_i \mathbf{P}_i^\alpha$: $\mathfrak{Q}$ has value q_i on α if and only if α has recording property $\mathcal{P}_i^\alpha$. Note that though Equation (3.1) constrains $\mathbf{H}_I$, this is a constraint which can be met: In particular, the arrow represents unitary evolution over the interval (t_0, t_1). Note also that any atomic system trivially counts as an M-type system, but no real apparatus is an atomic system.

To see why an interaction meeting these conditions is capable of being used to perform a measurement of the value of a quantum dynamical variable, consider the effect of such an interaction on a system σ when the initial universal system representative has the form

$$\psi^\omega(t_0) = \phi^\sigma \otimes \chi_0^\alpha \ , \text{ where } \phi^\sigma = \sum_{ij} c_{ij} \cdot \chi_{ij}^\sigma.$$

By the subspace decomposition condition, the initial system representative of σ is $[\phi^\sigma]$, and that of α is $[\chi_0^\alpha]$. By equation (3.1) in the definition of an SMI, and the linear evolution of the system representative of a noninteracting system, the final universal system representative is

$$\psi^\omega(t_1) = \sum_{ij} c_{ij} \cdot (\xi_{ij}^\sigma \otimes \chi_{ij}^\alpha).$$

As long as $(|c_{ij}|^2 = |c_{i'j'}|^2) \Rightarrow i = i'$, it follows from the subspace decomposition condition that the final system representatives of σ and α are the pair $([\xi_{ij}^\sigma], [\chi_{ij}^\alpha])$, for some i, where j varies over some index set I_{il} such that for all $j, j' \in I_{il}$, $|c_{ij}|^2 = |c_{ij'}|^2$. For later use, set $\cup_l I_{il} = I_i$. Consider first the case in which $c_{ij} = 0$ for $i \neq k$. Let $\mathcal{P}^\sigma(\mathcal{A} = a_k)$ be the property which σ has when $\mathcal{A} = a_k$. If $\mathcal{P}^\sigma(\mathcal{A} = a_k)$ is an irreducible property at t_0, then as long as the proviso is met, σ initially has

$\mathcal{P}^\sigma(\mathcal{A} = a_k)$ by the system representative condition. And since α is an M-type system, it follows that $\mathcal{P}_k^\alpha$ is prime, and that there is no composite property weaker than $\mathcal{P}_k^\alpha$ that α does not have; the system representative condition consequently implies that α has $\mathcal{P}_k^\alpha$ at t_1. Hence, if the initial universal system representative is $\psi^\omega(t_0)$, and $\mathcal{A}$ initially has value a_k on σ, then α is certain to acquire the recording property $\mathcal{P}_k^\alpha$ as a result of the interaction. Consider now the case in which there are distinct i, i' with $c_{ij} \neq 0$, $c_{i'j'} \neq 0$ (for some j, j'). Initially, σ does not have $\mathcal{P}^\sigma(\mathcal{A} = a_i)$ for any i. For since this is assumed to be an irreducible property of σ at t_0, σ could have it only by virtue of the system representative condition, which it does not have. Hence, $\mathcal{A}$ does not initially have a precise value on σ. Nevertheless, the final system representative of α is $[\chi_{ij}^\alpha$: fixed $i]$, with probability $\Sigma_{j \in I_{i}} |c_{ij}|^2$. Consequently, for some i, α has $\mathcal{P}_i^\alpha$ at t_1. Thus, in this case also α acquires some recording property $\mathcal{P}_i^\alpha$ as a result of the interaction: Indeed, the probability that it acquires $\mathcal{P}_i^\alpha$ is $\Sigma_{j \in I_{i}} |c_{ij}|^2$. This is why an SMI gives a definite result, even when the measured variable had no prior value.

Note that any dynamical variable $\mathcal{A}$ measurable in an SMI must correspond to a self-adjoint operator with pure point spectrum, since its eigenvectors span H^σ. Note also that a set of dynamical variables $\{\mathcal{A}_p\}$ measurable simultaneously in the same SMI must be compatible; that is, their corresponding self-adjoint operators must all commute pairwise, or, equivalently, they must all share a complete set of eigenvectors. To show this, consider the effect of an SMI for any pair $\mathcal{A}, \mathcal{B}$ of quantities in this set. Any initial universal system representative $\psi^{\sigma\alpha}(t_0)$ may be expanded in each of two ways:

$$\psi^{\sigma\alpha}(t_0) = \left(\sum_{ij} c_{ij}\chi_{ij}^\sigma \right) \otimes \chi_0^\alpha = \left(\sum_{kl} d_{kl}\zeta_{kl}^\sigma \right) \otimes \chi_0^\alpha ,$$

where $\mathbf{A}\chi_{ij}^\sigma = a_i\chi_{ij}^\sigma$, $\mathbf{B}\zeta_{kl}^\sigma = b_k\zeta_{kl}^\sigma$. The interaction therefore gives rise to a final universal system representative $\psi^{\sigma\alpha}(t_1)$ satisfying

$$\psi^{\sigma\alpha}(t_1) = \sum_{ij} c_{ij} \cdot (\xi^\sigma_{ij} \otimes \chi^\alpha_{ij}) = \sum_{kl} d_{kl} \cdot (\theta^\sigma_{kl} \otimes \eta^\alpha_{kl}).$$

But by the biorthogonal decomposition lemma, these two decompositions of $\psi^{\sigma\alpha}(t_1)$ must each define the same privileged sets of subspaces, whatever the initial system representative $\psi^{\sigma\alpha}(t_0)$. In particular, when the $|c_{ij}|^2$ are all nonzero and all unequal, they must each define the same privileged set of one-dimensional subspaces. Hence, it must be possible to relabel the second summation so that

$$\psi^{\sigma\alpha}(t_1) = \sum_{ij} c_{ij} \cdot (\xi^\sigma_{ij} \otimes \chi^\alpha_{ij}) = \sum_{ij} d_{ij} \cdot (\theta^\sigma_{ij} \otimes \eta^\alpha_{ij}) ,$$

with $|c_{ij}|^2 = |d_{ij}|^2$, $[\xi^\sigma_{ij}] = [\theta^\sigma_{ij}]$, and $[\chi^\alpha_{ij}] = [\eta^\alpha_{ij}]$. But the same interaction satisfies, for all i, both

$$\chi^\sigma_{ij} \otimes \chi^\alpha_0 \rightarrow \xi^\sigma_{ij} \otimes \chi^\alpha_{ij}$$

and also, after suitable relabeling,

$$\zeta^\sigma_{ij} \otimes \chi^\alpha_0 \rightarrow \theta^\sigma_{ij} \otimes \eta^\alpha_{ij}.$$

Hence, for all i, j, $[\zeta^\sigma_{ij}] = [\chi^\sigma_{ij}]$: therefore, **A,B** share a complete set of eigenvectors, and therefore commute. Thus, $\{\mathcal{A}_p\}$ constitutes a compatible set of dynamical variables.

Following von Neumann, measurement in quantum mechanics has often been thought to involve a sui generis discontinuous, indeterministic change in the quantum state of the object system, of the combined object-plus-apparatus system, or of both. According to the notorious **projection postulate**, the effect of a quantum measurement is to put the object system into a state in which its quantum state vector is an eigenvector of the measured variable, with eigenvalue equal to the measured value: And, at least in some versions, measurement simultaneously puts the apparatus system into an eigenstate of the recording quantity in which this has a value corresponding to the value of the measured quantity. An important aspect of the measurement problem, in its traditional form, is to reconcile the projection postulate with the time-dependent Schrödinger equation, according to which the quantum state vector of any isolated system, including an isolated object-plus-apparatus compound system, always evolves continuously and deterministically. This is not a problem in the present interpretation, which involves no projection postulate. However, among the many dubious reasons for requiring the projection postulate, there is one good reason which the present interpretation cannot afford to ignore. Whether or not immediate repetition of a quantum measurement on an "object" system always yields the same result (as the projection postulate requires it to do), it *is* true that quantum measurements are **verifiable**; that is, repeated observation of an "apparatus" system at the conclusion of a measurement interaction reveals it to be recording the same result of the measurement, as long as it is not reset, or otherwise grossly interfered with. This is a natural consequence of those versions of the projection postulate according to which measurement leaves the "apparatus" system in the appropriate eigenstate of the recording quantity: For in such an eigenstate it follows from the Born probability rules that a subsequent measurement of the value of the recording quantity will, with probability one, yield the corresponding eigenvalue. The question is, can the verifiability of measurement results be accounted for in the present interpretation *without* the projection postulate?

In order to answer this question, consider the nature of the interaction required to observe the recording property of "apparatus" system α after a measurement-type interaction between α and an "object" system σ which was suited to measure dynamical variable $\mathcal{A}$ on σ. This must be a measurement-type interaction between α (itself now regarded as an "object" system) and another "apparatus" system β. Suppose for simplicity that this is an SMI for recording quantity $\mathcal{Q}$ on α, and that it immediately follows an SMI for $\mathcal{A}$ on σ. Then the interaction Hamiltonian will be such as to induce the following evolution over the interval (t_1, t_2) of the new interaction

$$\chi_{ij}^\alpha \otimes \chi_0^\beta \rightarrow \xi_{ij}^\alpha \otimes \chi_{ij}^\beta \, ,$$

where the $\{\chi_{ij}^\alpha\}$ constitute a complete orthonormal set of eigenvectors of recording quantity $\mathfrak{Q}^\alpha$ ($\mathbf{Q}^\alpha \chi_{ij}^\alpha = q_i \chi_{ij}^\alpha$), and the recording property $\mathcal{P}_i^\beta$ records on system β the value q_i for $\mathfrak{Q}^\alpha$ on α. Now, if this observation of α is to be repeatable, it must not disturb the value q_i of $\mathfrak{Q}^\alpha$: Hence, it must leave α with eigenvalue q_i. Such an SMI will be said to be **minimally disturbing for $\mathfrak{Q}^\alpha$**. Given that this SMI is minimally disturbing for $\mathfrak{Q}^\alpha$, $\mathbf{Q}^\alpha \xi_{ij}^\alpha = q_i \xi_{ij}^\alpha$, for all j.

Now what will be the result of the measurement of $\mathfrak{Q}^\alpha$ on α by means of an SMI that is minimally disturbing for $\mathfrak{Q}^\alpha$? The foregoing analysis gives the following universal system representative at the start of this interaction, and just after an initial SMI for $\mathcal{A}$ on σ,

$$\psi^{\sigma\alpha\beta}(t_1) = \sum_{ij} c_{ij} \cdot (\xi_{ij}^\sigma \otimes \chi_{ij}^\alpha) \otimes \chi_0^\beta.$$

Assume the actual system representative of α is then $[\chi_{mj}^\alpha]$ ($j \in I_{mp}$), so that the actual value of $\mathfrak{Q}^\alpha$ at t_1 is q_m, recording the result a_m for the measurement of $\mathcal{A}$ on σ. The $\mathfrak{Q}^\alpha$-measurement interaction will then produce the following final universal system representative, assuming (for simplicity) it does not couple β to σ,

$$\psi^{\sigma\alpha\beta}(t_2) = \sum_{ij} c_{ij} \cdot \left(\mathbf{U}^\sigma(t_2 - t_1)\xi_{ij}^\sigma \otimes \xi_{ij}^\alpha \otimes \chi_{ij}^\beta \right).$$

Considering this first as a biorthogonal decomposition onto H^α and $\mathsf{H}^{\sigma\beta}$, note that the accessible subspaces for α at t_2 are of the form $[\xi_{ij}^\alpha]$ ($j \in I_{il}$), again assuming ($|c_{ij}|^2 = |c_{i'j'}|^2$) $\Rightarrow i = i'$. By the stability condition given in the previous section, since $\mathbf{Q}^\alpha \xi_{ij}^\alpha = q_i \xi_{ij}^\alpha$, the actual system representative of α at t_2 is $[\xi_{mj}^\alpha], j \in I_{ms}$ (for some s). Thus, the $\mathfrak{Q}^\alpha$ measurement interaction

91

has indeed not disturbed the value of $\mathcal{Q}^\alpha$. By the subspace decomposition condition, the system representative of $\sigma\oplus\beta$ at t_2 is $[\mathbf{U}^\sigma(t_2-t_1)\xi^\sigma_{mj} \otimes \chi^\beta_{mj}]$ ($j\in I_{ms}$). Now, considering the biorthogonal decomposition of $\psi^{\sigma\alpha\beta}(t_2)$ onto H^β and $\mathsf{H}^{\sigma\alpha}$, note that the accessible subspaces for β at t_2 are of the form $[\chi^\beta_{ij}]$ ($j\in I_{il}$). But given the system representative of $\sigma\oplus\beta$, all of these are ruled out by the minimal meshing condition except $[\chi^\beta_{mj}]$ ($j\in I_{ms}$). Also $\mathbf{P}^\beta_m\chi^\beta_{mj} = \chi^\beta_{mj}$, and since the $\mathcal{Q}^\alpha$ measurement interaction is an SMI for $\mathcal{Q}^\alpha$ on α, $\mathcal{P}^\beta_m$ is a prime property of β, and there are no weaker nontrivial composite properties of β. Hence, the proviso is met, and therefore by the system representative condition, β has $\mathcal{P}^\beta_m$ at t_2. It follows that a measurement of $\mathcal{Q}^\alpha$ on α by means of an SMI which is minimally disturbing for $\mathcal{Q}^\alpha$ not only leaves the value of $\mathcal{Q}^\alpha$ undisturbed, but is also certain to record this value of $\mathcal{Q}^\alpha$, namely, the value of $\mathcal{Q}^\alpha$ recording the previously measured value of $\mathcal{A}$ on σ.

It is now possible to explain why compound properties and prime properties are treated differently by the present interpretation, and specifically, why the composition condition, and the restriction of the system representative condition to prime properties, are necessary. Consider the properties $\mathcal{P}^\sigma$, $\mathcal{P}^\tau$, and $\mathcal{P}^\sigma\otimes\mathcal{P}^\tau$ (the property corresponding to the projection operator $\mathbf{P}^\sigma\otimes\mathbf{P}^\tau$). Associated with these properties are dynamical variables $\mathcal{R}^\sigma$, $\mathcal{R}^\tau$, and $\mathcal{R}^{\sigma\oplus\tau}$ defined as follows: $\mathcal{R}^\sigma$ ($\mathcal{R}^\tau$) has value 1 when σ has $\mathcal{P}^\sigma$ (τ has $\mathcal{P}^\tau$), and value 0 when σ has $\overline{\mathcal{P}}^\sigma$ (τ has $\overline{\mathcal{P}}^\tau$); $\mathcal{R}^{\sigma\oplus\tau}$ has value 1 when $\sigma\oplus\tau$ has $\mathcal{P}^\sigma\otimes\mathcal{P}^\tau$, and value 0 when $\sigma\oplus\tau$ has $\overline{\mathcal{P}^\sigma\otimes\mathcal{P}^\tau}$. Now consider SMIs for $\mathcal{R}^\sigma$ on σ and $\mathcal{R}^\tau$ on τ. The Hamiltonians for these interactions are such as to produce the following evolutions

$$\chi^\sigma_{dj} \otimes \mu^\sigma_0 \to \xi^\sigma_{dj} \otimes \mu^\sigma_{dj} \quad (d=0,1),$$

$$\lambda^\tau_{ek} \otimes \mu^\tau_0 \to \xi^\tau_{ek} \otimes \mu^\tau_{ek} \quad (e=0,1),$$

where μ^σ (μ^τ) is a vector in the Hilbert space of the "apparatus" system M^σ (M^τ) interacting with σ (τ). Therefore, a joint SMI for $\mathcal{R}^\sigma$ on σ and $\mathcal{R}^\tau$ on τ will have a Hamiltonian such that

$$\chi_{dj}^\sigma \otimes \chi_{ek}^\tau \otimes \mu_0^\sigma \otimes \mu_0^\tau \to \xi_{dj}^\sigma \otimes \xi_{ek}^\tau \otimes \mu_{dj}^\sigma \otimes \mu_{ek}^\tau \, .$$

But this Hamiltonian is M-suitable for $\mathcal{R}^{\sigma\oplus\tau}$ on $\sigma\oplus\tau$, with "apparatus" system $M^\sigma\oplus M^\tau$, which is M-ready at the start of the interaction. And if no distinction were drawn between prime and composite properties, the conditions (i) and (ii) could not be imposed on an M-type system in an SMI, and so this interaction would qualify as an SMI for $\mathcal{R}^{\sigma\oplus\tau}$ on $\sigma\oplus\tau$. Now $\sigma\oplus\tau$ will be observed to have $\mathcal{P}^\sigma\otimes\mathcal{P}^\tau$ just in case the result of this measurement of $\mathcal{R}^{\sigma\oplus\tau}$ is 1. But this is also a joint measurement of $\mathcal{R}^\sigma, \mathcal{R}^\tau$ which will have joint outcome (1,1) just in case $\mathcal{R}^{\sigma\oplus\tau}$ is measured to have value 1. And so $\sigma\oplus\tau$ will be observed to have $\mathcal{P}^\sigma\otimes\mathcal{P}^\tau$ in this interaction just in case σ is observed to have $\mathcal{P}^\sigma$ and τ is observed to have $\mathcal{P}^\tau$. It is true that other interactions are SMIs for $\mathcal{R}^{\sigma\oplus\tau}$ but not for $\mathcal{R}^\sigma, \mathcal{R}^\tau$. But each of these must yield the same result as this interaction if it is reliable. Hence, any observation of $\mathcal{P}^\sigma\otimes\mathcal{P}^\tau$ would give a positive outcome just in case a joint observation of $\mathcal{P}^\sigma, \mathcal{P}^\tau$ gives positive outcomes for both properties. This removes any empirical reason to deny that $\sigma\oplus\tau$ has $\mathcal{P}^\sigma\otimes\mathcal{P}^\tau$ if and only if σ has $\mathcal{P}^\sigma$ and τ has $\mathcal{P}^\tau$. But suppose that no distinction is drawn between prime and composite properties, the composition condition is abandoned, and the system representative condition (minus the proviso) applies to *all* dynamical properties. Then it may easily happen that $\sigma\oplus\tau$ does not have $\mathcal{P}^\sigma\otimes\mathcal{P}^\tau$, even though σ has $\mathcal{P}^\sigma$ and τ has $\mathcal{P}^\tau$. This would leave it quite mysterious as to why in this situation an observation of $\mathcal{P}^\sigma\otimes\mathcal{P}^\tau$ will give a positive outcome.

However, the composition condition removes this problem immediately, because if σ has $\mathcal{P}^\sigma$ and τ has $\mathcal{P}^\tau$, then $\sigma\oplus\tau$ must have $\mathcal{P}^\sigma\otimes\mathcal{P}^\tau$. It may still happen that $\sigma\oplus\tau$ has $\mathcal{P}^\sigma\otimes\mathcal{P}^\tau$ though σ does not have $\mathcal{P}^\sigma$ and/or τ does not have $\mathcal{P}^\tau$. This can occur if $\mathcal{P}^\sigma \otimes \mathcal{P}^\tau$ is weaker than some other composite property $\mathcal{Q}^{\sigma \oplus \tau}$ that $\sigma \oplus \tau$ has by the composition condition, even though $\mathcal{P}^\sigma, \mathcal{P}^\tau$ are not themselves weaker than any factors of $\mathcal{Q}^{\sigma \oplus \tau}$ possessed by σ and τ. If this does happen, then a joint SMI for $\mathcal{P}^\sigma, \mathcal{P}^\tau$ may not yield a joint positive outcome. But note that such a joint measurement is no longer an SMI for $\mathcal{P}^\sigma \otimes \mathcal{P}^\tau$, since it violates condition (i) on an M-type system: The supposed recording property corresponding to a positive outcome for $\mathcal{P}^\sigma \otimes \mathcal{P}^\tau$ is not prime, but composed of recording properties corresponding to positive outcomes for $\mathcal{P}^\sigma, \mathcal{P}^\tau$. There is no SMI which can yield a positive outcome for $\mathcal{P}^\sigma \otimes \mathcal{P}^\tau$ and a simultaneous negative outcome for $\mathcal{P}^\sigma$ or $\mathcal{P}^\tau$, since $\mathcal{P}^\sigma, \mathcal{P}^\tau$ and $\mathcal{P}^\sigma \otimes \mathcal{P}^\tau$ are not all simultaneously observable in any SMI. I conjecture that further investigation of the class of measurement-type interactions would reveal that $\mathcal{P}^\sigma, \mathcal{P}^\tau$ and $\mathcal{P}^\sigma \otimes \mathcal{P}^\tau$ are *never* simultaneously observable in any measurement-type interaction. If this conjecture is true, then even though it may happen that $\sigma \oplus \tau$ has $\mathcal{P}^\sigma \otimes \mathcal{P}^\tau$ while σ does not have $\mathcal{P}^\sigma$ and/or τ does not have $\mathcal{P}^\tau$, this is not a situation that can ever be *observed*.

§3.2 IDEALIZATIONS RELAXED

The considerations of §3.1 show that if there were any SMIs, then in certain apparently restricted circumstances an SMI might be used to perform a measurement of the value of a quantum dynamical variable. But one may reasonably wonder whether there are, or even could be, any SMIs: And even if one accepts them at least as a legitimate idealization of actually realizable interactions, one may doubt whether there are many situations in which an SMI (or something like it) would actually set up the required correlation in dynamical states of "object" and "apparatus" systems. These concerns will now be treated. The first worry is to be addressed by seeing how far the idealizations involved in the definition of an SMI may legitimately be relaxed. The second worry prompts one to probe the theoretical limits within which an SMI would function as required. I begin with this second concern.

In showing that an SMI can correlate the initial dynamical state of σ with the final dynamical state of α, three important simplifying assumptions were made: that $\sigma\oplus\alpha$ is itself the universal system ω; that the initial system representative of $\sigma\oplus\alpha$ is just the product of the initial system representatives of σ and of α; and that the initial system representative of σ is expressible as a linear superposition of eigenvectors of $\mathbf{A}$, all of whose complex coefficients have unequal moduli. The first of these assumptions is not justified in any actual measurement interaction, in which α is strongly coupled to its environment. No justification has yet been offered for either the second or the third assumption.

In fact, the second assumption may be justified as the natural expression of the reasonable requirement that in a well-conducted measurement, the initial states of the object and apparatus systems are separate and uncorrelated. If some un-suspected prior correlation were to obtain, it would not be surprising if a measurement failed to proceed as intended. But in the present interpretation, systems σ and α may be said to be **uncorrelated** just in case $\mathbf{P}^{\sigma\oplus\alpha} = \mathbf{P}^{\sigma} \otimes \mathbf{P}^{\alpha}$, where, for example, $\mathbf{P}^{\sigma}$ projects onto the system representative of σ. If σ and α are uncorrelated in this sense, then *all* of the properties of $\sigma\oplus\alpha$ reduce to properties of σ and of α, through the composition and weakening conditions, since $\sigma\oplus\alpha$ has property $\mathcal{P}^{\sigma} \otimes \mathcal{P}^{\alpha}$ (corresponding to the projection $\mathbf{P}^{\sigma} \otimes \mathbf{P}^{\alpha}$) by virtue of the composition condition, and then the weakening condition already determines that $\sigma\oplus\alpha$ has just the properties which would be attributed to it through the system representative condition applied to its representative $\mathbf{P}^{\sigma\oplus\alpha}$. Consequently, if σ and α are initially uncorrelated, as they should be at the start of any measurement, the initial system representative of $\sigma\oplus\alpha$ is just the product of those of σ and of α.

Since not only do σ and α interact with one another, but α is also likely to interact strongly with the environment, one cannot assume in any practical application of quantum mechanics to the circumstances of a measurement that $\sigma \oplus \alpha = \omega$. Rather, $\omega = \sigma \oplus \alpha \oplus \tau$, for some nonnull "environment" system τ. Suppose then that the initial universal system representative ψ^ω is given by the essentially unique biorthogonal expansion $\psi^\omega = \sum_k a_k \cdot (\eta_k^{\sigma \oplus \alpha} \otimes \eta_k^\tau)$. If σ and α are uncorrelated, the system representative of $\sigma \oplus \alpha$ is $[\eta_1^{\sigma \otimes \alpha}] = [\phi^\sigma \otimes \chi_0^\alpha]$, where the actual representatives of σ and α are $[\eta_1^\sigma] = [\phi^\sigma]$ and $[\eta_1^\alpha] = [\chi_0^\alpha]$, respectively. Assume further that σ and α are initially strongly uncorrelated; that is, every accessible representative $[\eta_k^{\sigma \oplus \alpha}]$ of $\sigma \oplus \alpha$ is of the form $[\eta_k^\sigma \otimes \eta_k^\alpha]$, for some accessible representatives $[\eta_k^\sigma], [\eta_k^\alpha]$ of σ, α. Then ψ^ω has the essentially unique triorthogonal expansion

$$\psi^\omega = \sum_k a_k \cdot (\eta_k^\sigma \otimes \eta_k^\alpha \otimes \eta_k^\tau),$$

where $\eta_1^\sigma = \phi^\sigma$, $\eta_1^\alpha = \chi_0^\alpha$ (for some $\chi_0^\alpha \in P_0^\alpha$).

Now consider the evolution of ψ^ω through the measurement interaction. So far, this interaction has been required to satisfy

$$\chi_{ij}^\sigma \otimes \chi_0^\alpha \to \xi_{ij}^\sigma \otimes \chi_{ij}^\alpha, \tag{3.1}$$

for $\chi_0^\alpha \in P_0^\alpha$. How would the interaction proceed from a different initial vector? It is natural to suppose that α is only sensitive to the interaction when it has system representative $[\chi_0^\alpha]$. This means that, for $\chi_k^\alpha \in P_0^{\alpha \perp}$, and arbitrary χ^σ,

$$\chi^\sigma \otimes \chi_k^\alpha \to \chi'^\sigma \otimes \chi_k'^\alpha, \tag{3.2}$$

where $\chi'^\sigma = U^\sigma \chi^\sigma$, $\chi_k'^\alpha = U^\alpha \chi_k^\alpha$, and U^σ, U^α are free evolution operators for σ and α, respectively. An evolution conforming to (3.1) and (3.2) is unitary provided that $(\chi_k'^\alpha, \chi_{ij}^\alpha) = 0$ (for all i, j, k). The interaction couples σ and α, but assume provisionally that it does not couple $\sigma \oplus \alpha$ to τ. Then ψ^ω evolves as follows, through the interaction (recall that $\eta_1^\sigma = \phi^\sigma = \sum_{ij} c_{ij} \cdot \chi_{ij}^\sigma$)

$$\psi^\omega \rightarrow \psi'^\omega =$$

$$\sum_{k \neq 1} a_k \cdot (\eta_k'^\sigma \otimes \eta_k'^\alpha \otimes \eta_k'^\tau) + a_1 \cdot \sum_{ij} c_{ij} \cdot (\xi_{ij}^\sigma \otimes \chi_{ij}^\alpha \otimes \eta_1'^\tau),$$

where $\eta_k'^\tau = \mathbf{U}^\tau \eta_k^\tau$, with $\mathbf{U}^\tau$ the free evolution operator for τ, so that $(\eta_k'^\tau, \eta_{k'}'^\tau) = (\eta_k^\tau, \eta_{k'}^\tau) = \delta_{kk'}$. Now this is a biorthogonal expansion onto H^α and $H^{\sigma \oplus \tau}$, and also onto H^τ and $H^{\sigma \oplus \alpha}$. Applying the subspace decomposition condition, the accessible representatives for α are $[\eta_k'^\alpha]$ ($k \neq 1$), $[\chi_{ij}^\alpha]$; and for τ, $[\eta_1'^\tau]$. But the *actual* initial system representative for α was $[\eta_1^\alpha] = [\chi_0^\alpha]$. Hence, the actual initial system representative for τ was $[\eta_1^\tau]$. Therefore, the final system representative for τ is $[\eta_1'^\tau]$; and so the actual final system representative for α is $[\chi_{ij}^\alpha]$, with probability $|c_{ij}|^2$. Now, in fact, α may well be strongly coupled to τ. But if α is to serve as a measuring device this coupling cannot be such as to disturb the value of the recording quantity on α. And the stability condition makes it clear how even a strong interaction between α and τ can preserve the value of the recording quantity while altering the system representative of α. Allowing for an interaction of this kind between α and τ, one can still conclude that the final system representative for α makes true $\mathfrak{Q} = q_i$ (for some i), with probability $\Sigma_j |c_{i'j}|^2$ that it makes true $\mathfrak{Q} = q_{i'}$. On the assumption that "object" and "apparatus" systems were initially strongly uncorrelated, the more realistic treatment has reproduced the key results of the earlier highly schematic account.

Consider, finally, what may be called the problem of degenerate final apparatus states. A set of complex coefficients Γ in an orthonormal expansion of a vector will be said to be **degenerate** coefficients just in case $|c_i|^2 = |c_j|^2$ for all c_i, c_j in Γ: An expansion with degenerate coefficients will be said to be degenerate. The universal system representative is expressible by a vector. That vector may be written as a biorthogonal expansion of vectors from two components $\sigma, \bar{\sigma}$ with complex coefficients. If this expansion is degenerate, the representatives of $\sigma, \bar{\sigma}$ will be multidimensional subspaces of their respective Hilbert spaces. This gives rise to a problem when one examines the effect of an SMI for $\mathscr{A}$ on a system σ whose representative is expressible by a degenerate expansion of ϕ^σ in terms of orthonormal eigenvectors of $\mathbf{A}$. It is true that for any natural measure of probability this situation will very probably not arise, since the condition that an expansion of ϕ^σ not be degenerate is that $(|c_{ij}|^2 = |c_{i'j'}|^2) \Rightarrow i = i', j = j'$. This is unlikely to fail, insofar as there are $2^{\aleph_0}$ possible values for each $|c_{ij}|^2$, but at most $\aleph_0$ distinct values of (i,j). But it is still important to consider whether there are any difficulties for the account of measurement interactions when the relevant expansion of ϕ^σ is degenerate.

The problematic case occurs when the degeneracy in the expansion of ϕ^σ is transferred to that of ψ'^ω, since there are distinct i, i' such that, for some j, j', $|c_{ij}|^2 = |c_{i'j'}|^2$. For in that case, neither $[\chi_{ij}^\alpha$: fixed $i]$ nor $[\chi_{i'j'}^\alpha$: fixed $i']$ will be an accessible final system representative for α. Only the subspace spanned by *all* vectors $\chi_{i'j'}^\alpha$ for which $|c_{i'j'}|^2 = |c_{ij}|^2$ will be accessible. Since this subspace is not included in any subspace $\mathbf{P}_i^\alpha$ corresponding to a "recording property," it follows that at the conclusion of the interaction the "apparatus" system α has no "recording property"; though α *does* have a property $\mathscr{P}_{\{i_1, i_2, \ldots\}}^\alpha$ corresponding to the subspace $\mathbf{P}_{i_1}^\alpha \oplus \mathbf{P}_{i_2}^\alpha \oplus \cdots$, where $\mathbf{P}_{i_1}^\alpha \oplus \mathbf{P}_{i_2}^\alpha \oplus \cdots$ includes the final system representative of α. In these circumstances, it appears that the apparatus will fail to indicate any definite result of the measurement: Its final state is degenerate. This is the analog within the present interpretation of one central aspect of the traditional measurement problem as it is presented by Wigner, for example. Though notice that the problem can now only arise for very special initial states of the measured system, and not for arbitrary nontrivial superposed initial states.

The answer to this problem is that once interactions between apparatus and environment are considered, the degeneracy which produces the difficulty will be lifted. Degenerate "final" dynamical states of the apparatus are metastable. External interactions with the environment would alter the final dynamical state of the apparatus so that if formerly it contained only the property $\mathcal{P}^{\alpha}_{\{i_1, i_2, \ldots\}}$, it would rapidly come to include some definite recording property (perhaps $\mathcal{P}^{\alpha}_{i_2}$). In practice, these external interactions prevent degenerate final states from being realized at all.

A simplified example should suffice to illustrate the removal of degeneracies. Thus, suppose that the initial system representative of σ is $\phi^{\sigma} = (1/\sqrt{2}) (\phi^{\sigma}_1 - \phi^{\sigma}_2)$, with $\{\phi^{\sigma}_1, \phi^{\sigma}_2\}$ a complete orthonormal basis for H^{σ}. Assume, for simplicity, that the initial universal system representative is $\psi^{\omega} = \phi^{\sigma} \otimes \chi^{\alpha}_0 \otimes \chi^{\tau}_0$. In the absence of "apparatus"-environment interactions, the final universal system representative would be

$$\psi^{\omega} = (1/\sqrt{2}) ((\xi^{\sigma}_1 \otimes \chi^{\alpha}_1 \otimes \chi'^{\tau}_1) - (\xi^{\sigma}_2 \otimes \chi^{\alpha}_2 \otimes \chi'^{\tau}_2)).$$

The final system representative for α would then be $[\chi^{\alpha}_1, \chi^{\alpha}_2]$, and so the "apparatus" would fail to indicate any definite result of the measurement. But now suppose that α interacts with the environment through an interaction which preserves the value of the recording quantity (as it must if this is to provide an enduring record of the outcome of the measurement)

$$\chi^{\alpha}_i \otimes \chi'^{\tau}_i \rightarrow \sum_j c_{ij} \cdot (\chi^{\alpha}_{ij} \otimes \chi'^{\tau}_{ij}) \quad (i = 1, 2),$$

where $Q\chi^{\alpha}_{ij} = Q\chi^{\alpha}_i$, and the right-hand side is a biorthogonal expansion. Then the universal system representative becomes

$$\psi'^{\omega} = (1/\sqrt{2}) \sum_j$$

$$\left(c_{1j} \cdot (\xi_1^{\sigma} \otimes \chi_{1j}^{\alpha} \otimes \chi'^{\tau}_{1j}) - c_{2j} \cdot (\xi_2^{\sigma} \otimes \chi_{2j}^{\alpha} \otimes \chi'^{\tau}_{2j}) \right).$$

For an arbitrary "apparatus"-environment interaction, there is no reason why, for any j, $|c_{1j}|^2 = |c_{2j}|^2$. Hence, there is no reason to expect this expansion to be degenerate, even though the expansion of ψ'^{ω} was degenerate. Therefore, a short time after the conclusion of the initial measurement-type interaction, the dynamical state of the "apparatus" will record a definite result for the measurement, even though, in the absence of any "apparatus"-environment interaction, the final "apparatus" dynamical state consistent with ψ'^{ω} does not.

There is nothing surprising about the need for an external disturbance to permit the correct functioning of a measurement device. A frictionless balance would never permit one to measure the weight of an object placed on it. And notice that the interaction with the environment is indeterministic, and in no way biases the result of the measurement it produces. There may still be a feeling that the result of a measurement should itself reflect the value the measured quantity actually had on the object system immediately beforehand (or would have had in the absence of the measurement). But it is characteristic of quantum measurements in the present interpretation, as well as for more orthodox interpretations, that they typically do not in this sense reveal possessed values.

So far, it has been assumed that SMIs occur, or at least that they might occur. I now wish to assess the significance of two different reasons for supposing that their occurrence would conflict with known features of interactions.

The first feature is the existence of conservation laws. There are well-known results due to Wigner (1952) and Araki and Yanase (1960) to the effect that no ideal measurement is possible of any quantum dynamical variable whose associated operator does not commute with all additively conserved

100

quantities. [Though Howard Stein has pointed out to me that there is actually a flaw in the argument of the latter paper, as further detailed in Stein and Shimony (1972).] While an interaction satisfying Equation (3.1) does not quite meet the conditions imposed on an ideal measurement by these authors, it is trivial to extend the proof of Araki and Yanase to apply to interactions satisfying Equation (3.1). Thus, suppose that a dynamical variable $\mathcal{L}$ with associated operator $\mathbf{L}$ (where $\mathbf{L} = \mathbf{L}^\sigma \otimes \mathbf{I}^\alpha + \mathbf{I}^\sigma \otimes \mathbf{L}^\alpha$) is conserved, in the sense that $[\mathbf{L}, \mathbf{U}_{t_0't_1}] = 0$, where $\mathbf{U}_{t_0't_1}$ is the evolution operator corresponding to Schrödinger evolution of a vector over the duration (t_0, t_1) of an interaction which is M-suitable for $\mathcal{A}$. It then follows that $[\mathbf{L}^\sigma, \mathbf{A}^\sigma] = 0$, as is shown in the appendix. Consequently, the class of variables $\mathcal{A}$ for which an M-suitable interaction is possible is restricted to those whose associated operators satisfy $[\mathbf{L}^\sigma, \mathbf{A}^\sigma] = 0$, for all additively conserved $\mathcal{L}$. This is a significant restriction: For no component of spin can there be an SMI which conserves any other component of spin, for example.

Now it is also well known that an *approximate* measurement may be made even of a quantity which does not commute with all additively conserved quantities, provided that α contains enough of each additively conserved quantity (see Araki and Yanase, 1960; Yanase, 1961). It is therefore natural to assume that an interaction which permits such an approximate measurement, though not strictly M-suitable for the measured quantity, is nevertheless a legitimate approximation to an M-suitable interaction. But this assumption is quite problematic in the present interpretation, as will now be shown. Let an **approximate SMI for** $\mathcal{A}$ satisfy not Equation (3.1) but Equation (3.3) over the interval (t_0, t_1)

$$\chi_{ij}^\sigma \otimes \chi_0^\alpha \rightarrow \delta(\xi_{ij}^\sigma \otimes \chi_{ij}^\alpha) + \epsilon(\eta_{ij}^\sigma \otimes \eta_{ij}^\alpha), \qquad (3.3)$$

where $(\chi_{ij}^\alpha, \eta_{i'j'}^\alpha) = 0$ for all i, j, i', j', $(\eta_{ij}^\alpha, \eta_{i'j'}^\alpha) = \delta_{ii'} \cdot \delta_{jj'}$, $(\eta_{ij}^\sigma, \xi_{ij}^\sigma) \neq 0$, $|\delta|^2 + |\epsilon|^2 = 1$, and $|\epsilon|^2 \ll 1$. For an initially superposed system representative of $\sigma \oplus \alpha$, this would yield

$$\left(\sum_{ij} c_{ij}\chi_{ij}^{\sigma} \right) \otimes \chi_0^{\alpha} \rightarrow \qquad (3.4)$$

$$\delta \sum_{ij} c_{ij} \left(\xi_{ij}^{\sigma} \otimes \chi_{ij}^{\alpha} \right) + \epsilon \sum_{ij} c_{ij} \left(\eta_{ij}^{\sigma} \otimes \eta_{ij}^{\alpha} \right).$$

The problem is that Equation (3.4) is no longer a biorthogonal decomposition, even if ϵ is made arbitrarily small. It appears that an approximate SMI for $\mathcal{A}$ will almost *always* leave α in a degenerate final state in which the recording quantity has no definite value.

Perhaps the best solution to this problem appeals to the fact that any actual quantum measurement involves a chain of amplification. From the present perspective, such a chain may be represented by a number of links, each consisting of an (approximate) M-type interaction which establishes a correlation between the dynamical state of one system and the resulting dynamical state of another. It is only necessary for some link in the chain to be a nonapproximate SMI for the quantity it measures in order for that interaction to lead to a nondegenerate final apparatus state. Suppose, for example, that α were subjected to an SMI at t_1 for a recording quantity $\mathcal{Q}$ with $\mathbf{Q}\chi_{ij}^{\alpha} = q_i\chi_{ij}^{\alpha}$, $\mathbf{Q}\eta_{ij}^{\alpha} = q_{-i}\eta_{ij}^{\alpha}$ ($i = 1, 2, \ldots$). This would produce the following evolution over (t_1, t_2):

$$\left[\delta \sum_{ij} c_{ij}(\xi_{ij}^{\sigma} \otimes \chi_{ij}^{\alpha}) + \epsilon \sum_{ij} c_{ij}(\eta_{ij}^{\sigma} \otimes \eta_{ij}^{\alpha}) \right] \otimes \chi_0^{\beta} \rightarrow$$

$$\left[\delta \sum_{ij} c_{ij}(\xi_{ij}^{\sigma} \otimes \xi_{ij}^{\alpha} \otimes \chi_{ij}^{\beta}) + \epsilon \sum_{ij} c_{ij}(\eta_{ij}^{\sigma} \otimes \eta_{ij}'^{\alpha} \otimes \eta_{ij}^{\beta}) \right]$$

But this is now a biorthogonal decomposition in the representation of $\sigma \oplus \alpha$ and of β. Hence, β will record a definite outcome, and for small enough ϵ, this will almost certainly correspond to a result of measuring $\mathcal{A}$ on σ (and not to a spurious value q_{-i} of $\mathcal{Q}$). But the recording quantity of an M-type interaction may be selected at will, and can therefore be chosen to commute with all additively conserved quantities. (This is made easier by the fact that if the recording quantity is to remain stable it must itself be conserved during subsequent interactions.) Though there may be few dynamical variables for which a single interaction is M-suitable, this provides no reason to doubt that for *any* dynamical variable there is a chain involving at least one SMI which can yield an arbitrarily accurate measurement of that dynamical variable.

The second reason why one might suppose that the occurrence of SMIs would conflict with known features of interactions is that few, if any, interactions cut off sharply after a finite time: Rather, an interaction dies away, approaching zero strength asymptotically. Now, the subspace decomposition condition was applied to show that the interactions involved in SMIs yield individual states of the "apparatus" in which the recording quantity has a definite value. But the subspace decomposition condition may only be applied directly to a system which is not interacting. Hence, the subspace decomposition condition cannot, in fact, be directly applied to show that after a finite, relatively short, time actual M-type interactions produce a definite recorded result.

Nevertheless, its indirect application here may be justified on the basis of the following considerations, which rest on plausible (though not as yet experimentally established) assumptions about the dynamics of the indeterministic transitions among dynamical states that (in the present interpretation) occur during any quantum measurement interaction. If a measurement-type interaction really had ceased, it would indeed follow from the subspace decomposition condition that the apparatus then records a definite result. But a relatively short time after the interaction reaches its peak, its strength will already have decreased sufficiently for there to be a very low rate of transition into the set of accessible dynamical "apparatus" states, and each of these states will also be very close to some final accessible dynamical "apparatus" state. The asymptotic applicability of the subspace decomposition condition thus requires that after even a relatively short time, observation of the "apparatus" is practically certain to reveal it to be recording a definite result of the measurement. And our observations of apparatus do not entitle us to make any stronger claim than this. Hence, the above account of measurement-type interactions indeed explains, or rather, explicates, the quantum mechanical predictions of the observed results of quantum measurements.

§3.3 THE ASSIGNMENT OF QUANTUM STATES

In the present interpretation, the instantaneous quantum state of a system provides a characterization of that system's current probabilistic dispositions, some of which will be manifested in a future interaction. This characterization is wholly objective – it need not represent anyone's *knowledge* of the system (though of course someone may come to know a system's quantum state). But it neither consists in, nor reduces to, a description of the system's current dynamical state. Rather, the current quantum state of a system σ is supervenient on the current dynamical states both of σ and of other systems which σ is a component, and also on the occurrence of an outcome in an *M*-type interaction that σ is to undergo. This makes quantum states a somewhat secondary feature of the quantum world, from an ontological point of view. Epistemologically, however, their role is primary: A central part of the empirical content of quantum mechanics is expressed by the Born rules, and these are stated in terms of the quantum state. Both our confidence in the empirical adequacy of quantum mechanics, and our ability to successfully apply the theory for a wide variety of predictive and explanatory purposes depend therefore on our ability to assign quantum states. How then is it possible to assign a quantum state to a system or ensemble of systems?

The first point that needs to be made in answer to this

question is that frequently it is *not* legitimate to assign either a pure or a mixed quantum state to a system or ensemble: Systems describable by quantum states are the exception rather than the rule. Only in rather special circumstances is it possible to characterize a system's probabilistic dispositions by the assignment of a quantum state. This should not seem surprising if one recalls that quantum states are in practice only assigned to systems that have been suitably prepared in some way, typically (though perhaps not exclusively) using some combination of prior interaction plus selection. In this section I limit myself to a consideration of how it is possible to assign quantum states to such systems.

The assignment of quantum states may be thought of as a device for facilitating probabilistic predictions concerning the outcome of *M*-type interactions. As such, it may be justified by demonstrating that predictions made using this device also follow from the basic probability relations of quantum mechanics, given in Chapter 2. Now, the usual Born rules give the general form for probabilistic predictions given an assignment of a quantum state. Hence, to explain and justify the practice of assigning quantum states it is both necessary and sufficient to derive the Born rules from the basic probabilistic relations of Chapter 2. Such a derivation presupposes a characterization of the class of *M*-type interactions, since the Born rules are taken to concern the final "apparatus" dynamical states following such interactions. §3.1 and §3.2 gave only a partial and preliminary characterization of this class, and so it will only be possible here to give a partial derivation of the Born rules. But even this partial derivation first requires an analysis of the conditions under which it is legitimate to assign a quantum state to a set of similar quantum systems, since only when these conditions are met will the Born rules apply.

There is another motive for investigating these conditions. Certain physical interactions are taken to prepare quantum states, in the sense that any collection of similar quantum systems subjected to these procedures is considered to have a given quantum state. The projection postulate provides a neat account of how it is possible to prepare quantum states, for those interpretations that include it, since, according to that postulate, a measurement will change the quantum state of the measured system, so that it becomes an eigenstate of the measured quantity with the measured eigenvalue. Hence, in order to prepare systems in a given quantum state, one measures a quantity for which that state is an eigenstate on each of a collection of similar systems, and selects just those systems that give the corresponding eigenvalue as the result of the measurement. These systems then have the desired quantum state. An interpretation such as the present one, which rejects the projection postulate, must account for state preparation in some other way. The key to the account given by the present interpretation is that a state preparation procedure does not discontinuously change the quantum state of systems to which it is applied: Rather, it becomes legitimate to assign a quantum state to a set of similar systems when they have been subjected to such a preparation procedure.

How can the class of state preparation procedures be characterized? In the present interpretation, this question presents a problem analogous to the problem of characterizing the class of measurement-type interactions. Its solution requires a similar strategy: Thus, as a first step, this section will analyze two particularly simple types of preparation procedure, with no claim that *all* state preparation procedures are of one of these types. There is a more direct relation to the earlier problem. Whether or not a given quantum state can be said to be prepared by a certain procedure itself depends on whether a subsequent M-type interaction involving systems subject to that procedure will in fact produce an outcome. From a different perspective, the entire procedure itself *includes* any M-type interactions which are undergone by those systems subjected to earlier stages of the procedure. And a quantum state is legitimately assigned not to systems that emerge from this entire procedure, but only to systems about to produce an outcome in the M-type interactions that (from this perspective) conclude the procedure. Conventionally, it is assumed that it is correct to assign a quantum state to an ensemble of suitably prepared systems irrespective of the outcome of any measurements on these systems. Although the present interpretation rejects this assumption, it is easy to explain why it seems legitimate. Most importantly, it follows from the Born rules that the value of the quantum state is wholly independent of the occurrence of *any* measurement. Secondly, it was assumed that any measurement yields a measured value: But as the present interpretation makes clear, some systems subjected to an M-type interaction produce no outcome – quantum states are assignable only to those systems which *do* produce an outcome. It is not incorrect to assign a quantum state to a system for which no measurement produces an outcome, as long as this is understood to have only counterfactual import: "If this system had produced an outcome in a measurement, then the probabilities of the various possible outcomes would have been given by the Born rules as applied to this quantum state."

A preparation procedure consists of an interaction followed by a selection according to the result of that interaction. Systems subject to the procedure undergo certain P-type physical interactions, which are (as usual) to be characterized quantum mechanically. But such an interaction may have one of a number of possible results, and the quantum state to be assigned to a system that undergoes the interaction depends on this result. For this reason, the preparation procedure also includes a selection of systems according to the results of the P-type interaction. It is important to note that this selection involves no further manipulation of the selected systems: *All* systems which undergo the P-type interaction are then subjected to the subsequent M-type interaction, whatever the result of the P-type interaction. It is equally important to note that systems for which the P-type interaction gives the appropriate result have the quantum state corresponding to that result, whether or not anyone is *aware* of the result, and, consequently, of their quantum state.

In the first type of simple preparation procedure, the "apparatus" system α which subsequently undergoes an M-type interaction with systems in the set to be assigned a quantum state is also involved in preparing that state. Let 'σ' denote a type of quantum system. A **type-1 simple preparation procedure (SPP) for state ξ^σ** consists of two elements.

i.   The first element is a P-type interaction among each system of a set Σ of systems of type σ, system α, and the environment ϵ, characterized by a Hamiltonian such that

$$\chi_i^\sigma \otimes \chi_{00}^\alpha \otimes \chi_0^\epsilon \rightarrow \xi_i^\sigma \otimes \chi_{i0}^\alpha \otimes \chi_i^\epsilon$$

for every $\chi_{00}^\alpha \in P_{00}^\alpha$, $\chi_0^\epsilon \in P_0^\epsilon$, where $\xi^\sigma = \xi_m^\sigma$ (for some m). [Strictly speaking, each system of type σ interacts with a *different* pair of α and ϵ systems – in a more realistic model of a preparation procedure, the same α, ϵ would be used for each σ-type system, but different (though analogous) properties would be observed on α for each individual interaction with a σ-type system.] This interaction is required to be an SMI for a complete, commuting set of dynamical variables on each system of type σ, accompanied by a further interaction coupling $\sigma \oplus \alpha$ to its environment ϵ. The χ_i^ϵ satisfy $(\chi_i^\epsilon, \chi_j^\epsilon) = \delta_{ij}$.

ii.   The second element is a selection of a subset Σ' of Σ such that, for each system in Σ', the property $\mathscr{P}_m^\alpha$ is possessed by α at the conclusion of the interaction, where $\mathbf{P}_m^\alpha \chi_{i0}^\alpha = \delta_{mi} \chi_{i0}^\alpha$.

Suppose that a set Σ' of systems has been assigned quantum state ξ^σ following a type-1 SPP for ξ^σ, and that each system in Σ is now subjected to an SMI for dynamical variable $\mathscr{B}$,

108

with "apparatus" system α. The Hamiltonian is such as to produce the evolution

$$\zeta_{jk}^\sigma \otimes \chi_{i0}^\alpha \to \xi_{jk}^\sigma \otimes \chi_{ijk}^\alpha,$$

for $\chi_{i0}^\alpha \in P_{i0}^\alpha$, where $B\zeta_{jk}^\sigma = b_j\zeta_{jk}^\sigma$, $\xi_i^\sigma = \sum_{jk} d_{ijk}\zeta_{jk}^\sigma$ ($\sum_{jk} |d_{ijk}|^2 = 1$), $\{\zeta_{jk}^\sigma\}$ is a complete orthonormal basis for H^σ and $(\chi_{ijk}^\alpha, \chi_{i'j'k'}^\alpha) = \delta_{ii'} \cdot \delta_{jj'} \cdot \delta_{kk'}$. If the initial universal system representative were

$$\psi^{\sigma\alpha\epsilon}(t_0) = \left(\sum_i c_i\chi_i^\sigma\right) \otimes \chi_{00}^\alpha \otimes \chi_0^\epsilon,$$

then, prior to the second SMI,

$$\psi^{\sigma\alpha\epsilon}(t_1) = \sum_i c_i \cdot \left(\xi_i^\sigma \otimes \chi_{i0}^\alpha \otimes \chi_i^\epsilon\right)$$

$$= \sum_{ijk} c_i d_{ijk} \cdot \left(\zeta_{jk}^\sigma \otimes \chi_{i0}^\alpha \otimes \chi_i^\epsilon\right)$$

and, subsequent to the second SMI,

$$\psi^{\sigma\alpha\epsilon}(t_2) = \sum_{ijk} c_i d_{ijk} \cdot \left(\xi_{jk}'^\sigma \otimes \chi_{ijk}^\alpha \otimes \chi_i'^\epsilon\right),$$

where $\chi_i'^\epsilon = U^\epsilon(t_2 - t_1)\chi_i^\epsilon$, from which it follows that $(\chi_i'^\epsilon, \chi_j'^\epsilon) = \delta_{ij}$. Considering this as a biorthogonal decomposition onto $H^\alpha, H^{\sigma\oplus\epsilon}$, the accessible system representatives for α following the second SMI are $[\chi_{ijk}^\alpha$: fixed $i,j]$, with k varying over some index set I_{ijl} such that $k,k' \in I_{ijl}$ if and only if $|d_{ijk}|^2 = |d_{ijk'}|^2$. (For later use, set $I_{ij} = \cup_l I_{ijl}$.) Moreover, each has probability $\sum_{k \in I_{ijl}} |c_i d_{ijk}|^2$. By the system representative condition, in the ijlth subspace, α has $\mathscr{P}_i^\alpha$, and also $\mathscr{P}^\alpha(\mathscr{B} = b_j)$. Moreover, by the stability condition, α has $\mathscr{P}_i^\alpha$ at t_2 just in case α has $\mathscr{P}_i^\alpha$ at t_1. Hence, the probability that α has $\mathscr{P}_m^\alpha$ at t_1 and $\mathscr{P}^\alpha(\mathscr{B} = b_j)$ at t_2 equals $\sum_{k \in I_{mj}} |c_m d_{mjk}|^2$. Therefore, the conditional probability that α has $\mathcal{P}^\alpha(\mathcal{B} = b_j)$ at t_2, given that α has $\mathcal{P}_m^\alpha$ at t_1, is given by

$$\text{prob}\left(\mathcal{P}^\alpha(\mathcal{B} = b_j) \text{ at } t_2 \;\middle|\; \mathcal{P}_m^\alpha \text{ at } t_1\right) = \frac{\Sigma_k |c_m d_{mjk}|^2}{|c_m|^2}.$$

Now the quantum state ξ^σ is assigned to the set Σ' of systems of type σ for which α has the property $\mathcal{P}_m^\alpha$ at t_1. Hence, for this set, the probability that α has $\mathcal{P}^\alpha(\mathcal{B} = b_j)$ at t_2 is just

$$\frac{\Sigma_k |c_m d_{mjk}|^2}{|c_m|^2} = \sum_{k \in I_{mj}} |d_{mjk}|^2,$$

and this is exactly the probability specified by the Born rules for observing value b_j in a measurement of $\mathcal{B}$ on each of a collection of systems of type σ in quantum state ξ^σ. Thus, if state ξ^σ is prepared by means of a type-1 SPP using α, and a measurement of any dynamical variable $\mathcal{B}$ is subsequently made in an SMI for $\mathcal{B}$, again using α, then the Born rules correctly give the probabilities for the various results of the $\mathcal{B}$-measurement. Note that if a P-type interaction satisfies $\xi_i^\sigma = \chi_i^\sigma$, then the state assigned after the associated type-1 SPP will be a simultaneous eigenstate of the complete commuting set of dynamical variables for which this interaction is an SMI. In that case it will be justifiable to apply the projection postulate after a type-1 SPP.

The second type of simple preparation procedure may be used to prepare the internal quantum state of a set of quantum systems by coupling their internal state both to their external state and to their environment. Though it corresponds more closely to many laboratory preparation procedures than does the first type, it is subject to theoretical restrictions which make it, in a sense, less general.

Suppose that the Hilbert space H^σ of a quantum system of type σ is of the form $H^\sigma = H^{\sigma I} \otimes H^{\sigma E}$, where the internal state of such a system is represented in $H^{\sigma I}$, and its external "spatial" state in $H^{\sigma E}$. Let $\{\chi_i^{\sigma I}\}$ be a complete orthonormal

110

basis for $H^{\sigma I}$. In a type-2 SPP for internal state $\xi^{\sigma I}$, it is *assumed* that it is legitimate to assign a quantum state vector $\phi_0^{\sigma E}$ to represent the external state of a set Σ of systems of type σ: $\phi_0^{\sigma E}$ might, for example, represent a wave packet centered on a certain z-momentum and a certain position in the x,y-plane, characterizing a molecular beam. To prepare an internal quantum state for a subset of Σ, the systems in Σ are subjected to an interaction that couples their internal state both to their environment, and to their external state: In a sense, the external dynamical state of these systems then serves as the dynamical state of a hypothetical "apparatus system," which records the result of the P-type interaction and permits the selection required by the SPP. More specifically, a **type-2 simple preparation procedure for internal state $\xi^{\sigma I}$** consists of two elements.

i.   The first element is a P-type interaction between each of a set Σ of systems of type σ, and the environment ϵ, characterized by a Hamiltonian such that

$$(\chi_i^{\sigma I} \otimes \phi_{00}^{\sigma E}) \otimes \chi_0^{\epsilon} \rightarrow (\xi_i^{\sigma I} \otimes \phi_{i0}^{\sigma E}) \otimes \chi_i^{\epsilon}$$

for every $\phi_{00}^{\sigma E} \in P_{00}^{\sigma E}$, $\chi_0^{\epsilon} \in P_0^{\epsilon}$, where $\xi^{\sigma I} = \xi_m^{\sigma I}$ (for some m), and the χ_i^{ϵ} satisfy $(\chi_i^{\epsilon}, \chi_j^{\epsilon}) = \delta_{ij}$. This interaction is analogous to an SMI for a complete, commuting set of *internal* dynamical variables on each system of type σ, accompanied by a further interaction coupling σ to ϵ: The only difference is that the interaction couples the internal state of the σ-type systems to their external state, instead of coupling their entire state to the state of a separate "apparatus" system.

ii. A selection of a subset Σ' of Σ such that each system in Σ' has the property $\mathcal{P}_m^{\sigma E}$ at the conclusion of the P-type interaction, where $\mathbf{P}_m^{\sigma E} \phi_{i0}^{\sigma} = \delta_{mi} \cdot \phi_{i0}^{\sigma}$. (Note that since σ-type systems will typically be microscopic, any observation of this property will be quite indirect. Indeed, as we shall see, the actual observation and selection based on it is only made *after* the measurement for which the state $\xi^{\sigma I}$ is prepared.)

Suppose that a set Σ' of systems has been assigned internal state $\xi^{\sigma I}$ following a type-2 SPP for internal state $\xi^{\sigma I}$, and that each system in Σ is now subjected to another interaction that couples its internal state to its external state in a way analogous to an SMI for internal dynamical variable $\mathcal{B}$, and simultaneously also couples the σ-type systems once more to their environment. Suppose the Hamiltonian is such as to produce the evolution

$$\zeta^{\sigma I}_{jk} \otimes \phi^{\sigma E}_{i0} \otimes \chi^{\epsilon}_{i} \to \xi'^{\sigma I}_{jk} \otimes \phi^{\sigma E}_{ijk} \otimes \chi^{\epsilon}_{ijk}$$

for $\phi^{\sigma E}_{i0} \in P^{\sigma E}_{i0}$, where

$$\mathbf{B}\zeta^{\sigma I}_{jk} = b_j \zeta^{\sigma I}_{jk}, \quad \xi^{\sigma I}_i = \sum_{jk} d_{ijk} \zeta^{\sigma I}_{jk} (\sum_{jk} |d_{ijk}|^2 = 1), \quad \{\zeta^{\sigma I}_{jk}\}$$

is a complete orthonormal basis for $H^{\sigma I}$, $(\chi^{\alpha}_{ijk}, \chi^{\alpha}_{i'j'k'}) = \delta_{ii'} \cdot \delta_{jj'} \cdot \delta_{kk'}$, and $(\chi^{\epsilon}_{ijk}, \chi^{\epsilon}_{i'j'k'}) = \delta_{ii'} \cdot \delta_{jj'} \cdot \delta_{kk'}$.

If the initial universal system representative were

$$\psi^{\sigma\epsilon}(t_0) = \left(\left(\sum_i c_i \chi^{\sigma I}_i \right) \otimes \phi^{\sigma E}_{00} \right) \otimes \chi^{\epsilon}_0 ,$$

then, prior to the second interaction,

$$\psi^{\sigma\epsilon}(t_1) = \sum_i c_i \left((\xi^{\sigma I}_i \otimes \phi^{\sigma I}_{i0}) \otimes \chi^{\epsilon}_i \right)$$

$$\psi^{\sigma\epsilon}(t_1) = \sum_{ijk} c_i d_{ijk} \left((\zeta^{\sigma I}_{jk} \otimes \phi^{\sigma E}_{i0}) \otimes \chi^{\epsilon}_i \right)$$

and, subsequent to the second SMI,

$$\psi^{\sigma\epsilon}(t_2) = \sum_{ijk} c_i d_{ijk} \left((\xi'^{\sigma I}_{jk} \otimes \phi^{\sigma E}_{ijk}) \otimes \chi^{\epsilon}_{ijk} \right)$$

This is a biorthogonal expansion onto H^{σ} and H^{ϵ}, and the accessible system representatives for σ are $[\xi'^{\sigma I}_{jk} \otimes \phi^{\sigma E}_{ijk} : $ fixed

112

i,j], with k varying over some index set I_{ijl} such that $k,k' \in I_{ijl}$ if and only if $|d_{ijk}|^2 = |d_{ijk'}|^2$. (For later use, set $I_{ij} = \cup_l I_{ijl}$.) Moreover, each has probability $\Sigma_{k \in I_{ijl}} |c_i d_{ijk}|^2$. Let $\mathscr{P}^\sigma_{mj}$ be the property of σ corresponding to the projection operator $\mathbf{I}^{\sigma I} \otimes \mathbf{P}^{\sigma E}_{mj}$, where $\mathbf{P}^{\sigma E}_{mj}$ projects onto the subspace $[\phi^{\sigma E}_{mjk} :$ fixed $m,j]$ ($k \in I_{mj}$) of $\mathsf{H}^{\sigma E}$. Since, for $j \neq j'$, $[\mathbf{P}^{\sigma E}_{mj}, \mathbf{P}^{\sigma E}_{mj'}] = \mathbf{0}$, it is possible simultaneously to indirectly observe whether or not σ has $\mathscr{P}^\sigma_{mj}$, for fixed m but variable j. By the system representative condition, a system from Σ will have $\mathscr{P}^\sigma_{mj}$ at t_2 just in case its system representative is $[\xi'^{\sigma I}_{mk} \otimes \phi^{\sigma E}_{mjk} :$ fixed $m,j]$ ($k \in I_{mjl}$, for some l). The set Σ' to be assigned internal quantum state $\xi^{\sigma I}$ at t_1 consists of that subset of Σ which has $\mathscr{P}^\sigma_{mj}$ (for some j) at t_2. In that set, the probability of $\mathscr{P}^\sigma_{mj}$ at t_2 is

$$\mathrm{prob}\left(\mathscr{P}^\sigma_{mj} \text{ at } t_2 \,\middle|\, \mathscr{P}^\sigma_{mj} \text{ (for some } j\text{) at } t_2 \right)$$

$$= \frac{\displaystyle\sum_k |c_m d_{mjk}|^2}{\displaystyle\sum_{jk} |c_m d_{mjk}|^2} = \sum_{k \in I_{mj}} |d_{mjk}|^2.$$

But this is exactly the probability specified by the Born rules for observing the value b_j of $\mathscr{B}$ on each of a collection of systems of type σ with internal quantum state $\xi^{\sigma I}$. Thus, if an internal quantum state $\xi^{\sigma I}$ of a set of systems of type σ is prepared by means of a type-2 SPP for $\xi^{\sigma I}$, and a subsequent observation of their correlated external dynamical state is taken to constitute a measurement of the internal dynamical variable $\mathscr{B}$ on systems of type σ, then the Born rules correctly give the probabilities for the various possible results of this $\mathscr{B}$-measurement. Note that passage of a beam through two Stern-Gerlach devices oriented at an angle to one another provides a good example of a type-2 SPP, for a quantum state which is an eigenstate of spin–component along the axis of the first device, followed by a measurement of spin-component along the axis of the second device. In this example, and in others, the interaction involved satisfies $\xi_i^{\sigma l} = \chi_i^{\sigma l}$. In such a case, the internal quantum state assigned after the type-2 SPP will be a simultaneous eigenstate of the complete commuting set of *internal* dynamical variables for which this interaction is an analog of an SMI. It will then be justifiable to apply the projection postulate to the internal state of systems subjected to a type-2 SPP.

In both these examples of simple preparation procedures, the quantum state assigned was a pure state – a vector which in fact coincided with the system representative. But the quantum state assigned to a system (or ensemble) need not be represented by a vector; and even when it is, this need not coincide with the system representative. An example in which the quantum state is quite distinct from the system representative will be given in Chapter 4. This example is also interesting because it illustrates the fact that, in the present interpretation, a system may be certain to manifest some dynamical property upon measurement, even though it does not possess that property. The property in question is not contained in the system's dynamical state, even though it is assigned probability 1 by the system's quantum state!

More typically, when a system has a quantum state, this will coincide with its system representative. Since a system representative may be a multidimensional subspace, the fundamental mathematical representative of a quantum state, in the present interpretation, is neither a vector nor a general density operator, but rather a subspace of vectors (or the corresponding projection operator). Is this a defect of the present interpretation? It seems to me that it is not. When normalized, a projection onto a finite dimensional subspace is just a density operator. And in the present interpretation, it is easy to understand why one conventionally represents quantum states either by vectors or by density operators, even though the fundamental mathematical representative of a quantum state is a subspace of vectors.

4

Coupled systems

In this and the following chapter I apply the present interpretation of quantum mechanics to coupled systems of the
type studied by Einstein et al. (1935), Bohm (1951), Bell
(1964), and Aspect et al. (1982a, 1982b) (to name just a few
seminal contributors to the large and growing literature on
this topic). The main conclusions were already anticipated in
Chapter 1. In the present interpretation, quantum mechanics
offers a detailed theoretical account of the physical events and
processes that underlie the correlations exhibited in measurements on such systems. This account depends both on
the postulation of irreducible properties of compound systems, and also on the treatment of measurement as a quantum
mechanical interaction. In the present chapter I present technical details of this account, and in the next chapter I explore
its metaphysical aspects and implications, especially for the
notions of holism and causal explanation.

For concreteness, I restrict attention here to systems composed of two spin-½ atomic systems: The generalization to
other pairs of systems with different spins is straightforward.
Such systems may result from a prior interaction between
the two component systems, or they may be produced de
novo (as a positron–electron pair may be produced by annihilation of a γ-ray). Call the component systems A, B and
the compound system $A \oplus B$. The system representative of
$A \oplus B$ will be a subspace of the tensor product Hilbert space
$H^A \otimes H^B$. In the case corresponding to the singlet spin state,
this is spanned by a vector which does not have a unique
biorthogonal decomposition in the component systems' Hil-

116

bert spaces, and in the present interpretation this complicates the treatment of the original Bohm version of the EPR thought-experiment. Consequently, in a presentation of an explanatory account of EPR-type correlations it will be best to work with a more general superposition of the singlet and triplet spin states. Assume then that A, B are two spin-½ systems which are no longer interacting (the interaction term in their joint Hamiltonian is zero); and that the initial compound system representative corresponds to a spin state which is a superposition of the singlet and triplet states, and an eigenvector of z-component of total spin, namely,

$$\psi^{A\oplus B} = \cos\theta \cdot (\zeta_+^A \otimes \zeta_-^B) - \sin\theta \cdot (\zeta_-^A \otimes \zeta_+^B),$$

where $\zeta_\pm^i$ is an eigenvector of z-component of spin for the ith system ($i = A,B$) with eigenvalue $\pm \hbar/2$. (Note that setting $\theta = \pi/4$ here gives the original Bohm version of the EPR thought-experiment.) The problem is to account for the correlations in the observed spin-components of systems A and B. Consider, then, two measurements M_A and M_B, on systems A and B, respectively: M_B is a measurement of the component of spin of system B along an axis making an angle 2ϕ with the z-axis, and M_A is a measurement of the component of spin of system A along an axis making an angle 2ψ with the z-axis, where $\tan\psi = -\tan\theta/\tan\phi$. If, for example, $\phi = \pi/4$, then $\psi = \pi - \theta$, and the axes of the M_A and M_B measurements are as shown in Figure 4.1 in the x,z-plane. If $\phi = \pi/3$, then $\tan\psi = -1/\sqrt{3} \cdot \tan\theta$, and the axes are as shown in Figure 4.2, with $2\pi - 2\psi < 2\theta$. If the M_B measurement gives the result $+\hbar/2$ (spin-up), then the M_A measurement is also certain to give the result $+\hbar/2$, as will now be shown.

Note first that the eigenvector ϕ_+ for spin-up along the axis of the M_B measurement may be obtained by applying a rotation through an angle 2ϕ about the y-axis to ζ_+. The rotation operator $\mathbf{R}_y(2\phi)$ is given by $e^{-i\phi\sigma_y}$, where σ_y is the

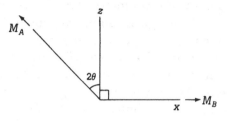

Figure 4.1

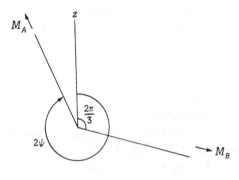

Figure 4.2

Pauli spin operator. Working in a representation in which σ_z is diagonal,

$$\zeta_+ = \begin{pmatrix} 1 \\ 0 \end{pmatrix}, \ \zeta_- = \begin{pmatrix} 0 \\ 1 \end{pmatrix}, \ \sigma_y = \begin{pmatrix} 0 & -i \\ i & 0 \end{pmatrix}.$$

Hence,

$$\phi_+ = e^{-i\phi\sigma_y} \cdot \zeta_+ = \left[\cos\phi \begin{pmatrix} 1 & 0 \\ 0 & 1 \end{pmatrix} - i\sin\phi \begin{pmatrix} 0 & -i \\ i & 0 \end{pmatrix} \right] \cdot \begin{pmatrix} 1 \\ 0 \end{pmatrix}$$

$$= \begin{pmatrix} \cos\phi & -\sin\phi \\ \sin\phi & \cos\phi \end{pmatrix} \cdot \begin{pmatrix} 1 \\ 0 \end{pmatrix} = \begin{pmatrix} \cos\phi \\ \sin\phi \end{pmatrix}$$

$$= \cos\phi \cdot \zeta_+ + \sin\phi \cdot \zeta_-.$$

Similarly, $\phi_- = -\sin\phi \cdot \zeta_+ + \cos\phi \cdot \zeta_-.$

Hence,

$$\zeta_+ = \cos\phi \cdot \phi_+ - \sin\phi \cdot \phi_-, \quad \zeta_- = \sin\phi \cdot \phi_+ + \cos\phi \cdot \phi_-.$$

And so, $\psi^{A \oplus B}$ may be written as

$$
\begin{aligned}
\psi^{A \oplus B} &= \cos\theta \cdot \zeta_+^A \otimes (\sin\phi \cdot \phi_+^B + \cos\phi \cdot \phi_-^B) \\
&\quad - \sin\theta \cdot \zeta_-^A \otimes (\cos\phi \cdot \phi_+^B - \sin\phi \cdot \phi_-^B) \\
&= (\cos\theta\sin\phi \cdot \zeta_+^A - \sin\theta\cos\phi \cdot \zeta_-^A) \otimes \phi_+^B \\
&\quad + (\cos\theta\cos\phi \cdot \zeta_+^A + \sin\theta\sin\phi \cdot \zeta_-^A) \otimes \phi_-^B ,
\end{aligned}
$$

which is of the form

$$
\begin{aligned}
\psi^{A \oplus B} &= N_+ (\cos\psi \cdot \zeta_+^A + \sin\psi \cdot \zeta_-^A) \otimes \phi_+^B \\
&\quad + N_- (\cos\chi \cdot \zeta_+^A + \sin\chi \cdot \zeta_-^A) \otimes \phi_-^B ,
\end{aligned}
$$

with

$$
\begin{aligned}
N_+\cos\psi &= \cos\theta\sin\phi, & N_-\cos\chi &= \cos\theta\cos\phi, \\
N_+\sin\psi &= -\sin\theta\cos\phi, & N_-\sin\chi &= \sin\theta\sin\phi.
\end{aligned}
$$

This gives

$$\tan\psi = -\tan\theta/\tan\phi, \quad \tan\chi = \tan\theta\tan\phi,$$

$$N_+ = \pm(\cos^2\theta\sin^2\phi + \sin^2\theta\cos^2\phi)^{1/2},$$
$$N_- = \pm(\cos^2\theta\cos^2\phi + \sin^2\theta\sin^2\phi)^{1/2}.$$

But $\psi_+ =_{df} \cos\psi \cdot \zeta_+ + \sin\psi \cdot \zeta_-$ is a spin-up eigenvector of spin-component in a direction in the x,z-plane making an angle 2ψ with the z-axis, and $\chi_+ =_{df} \cos\chi \cdot \zeta_+ + \sin\chi \cdot \zeta_-$ is a spin-up eigenvector of spin-component in a direction in the x,z-plane making an angle 2χ with the z-axis.

Now the probability that the M_B measurement gives result spin-up is just N_+^2, whereas the probability that the M_B measurement gives spin-up *and* the M_A measurement gives spin-up is also N_+^2. Consequently, the conditional probability, that the M_A measurement gives spin-up *given* that the M_B measurement gives spin-up, equals 1. Hence, if the M_B measurement gives the result spin-up, then the M_A measurement is also certain to give spin-up. Note that the original Bohm version of the EPR thought-experiment is recovered by setting $\theta = \pi/4$. Then $\psi = \phi + \pi/2$, $\chi = \phi$, the M_A and M_B measurements are performed along axes lying in opposite directions in space, and we get the usual perfect anticorrelations for spin measurements on both systems of the spin along the same arbitrary axis in the x,z-plane. Note also that setting $\phi = \pi/4$ gives $\psi = \pi - \theta$ and $\chi = \theta$, and $\psi^{A \oplus B}$ then has the following form

$$\psi_{\pi/4}^{AB} = (1/\sqrt{2}) \left[(\delta^A \otimes \zeta_+^B) - (\epsilon^A \otimes \zeta_-^B) \right],$$

where $\delta = \cos\theta \cdot \zeta_+ - \sin\theta \cdot \zeta_-$, $\epsilon = \cos\theta \cdot \zeta_+ + \sin\theta \cdot \zeta_-$, and $\xi_\pm = (1/\sqrt{2})(\zeta_+ \pm \zeta_-)$. This case, in which the M_B measurement is along the x-axis, and the M_A measurement along an axis in the x,z-plane making an angle 2θ with the z-axis (on the opposite side to the x-axis), was introduced by Kochen in an earlier (unpublished) treatment of the EPR correlations. Note however that $|N_+|^2 = |N_-|^2$ for the case $\phi = \pi/4$ (in fact, $N_+ = -N_-$). It will be seen shortly that this makes the biorthogonal expansions of certain multiple tensor product vectors nonunique, and thus complicates the analysis of the interactions involved in the M_A, M_B measurements. That is why the more general analysis (with general θ, ϕ) is required in the account of EPR-type correlations for the present interpretation.

To understand the origins of the correlations it is necessary to examine the effects of the M_A and M_B interactions on the properties of A, B, and $A \oplus B$. Prior to both interactions, the spin part of the system representative of $A \oplus B$ is spanned by the vector $\psi^{A \oplus B}$

$$\psi^{A \oplus B} = \cos\theta \cdot (\zeta_+^A \otimes \zeta_-^B) - \sin\theta \cdot (\zeta_-^A \otimes \zeta_+^B).$$

Provided that $\theta \neq (2n+1)\pi/4$ ($n = 0, 1, \ldots$), this represents an essentially unique biorthogonal expansion, and in accordance with the subspace decomposition condition, the possible system representatives for (A,B) are $([\zeta_+^A], [\zeta_-^B])$ or $([\zeta_-^A], [\zeta_+^B])$, with probabilities $(\cos^2\theta, \sin^2\theta)$. Hence, in accordance with the system representative condition, systems (A, B) have one of the following pairs of properties $(S_z^A(\text{up}), S_z^B(\text{down}))$ or $(S_z^A(\text{down}), S_z^B(\text{up}))$; and in accordance with the composition condition, $A \oplus B$ has one of the corresponding properties $(S_z^A(\text{up}) \wedge S_z^B(\text{down}))$ or $(S_z^A(\text{down}) \wedge S_z^B(\text{up}))$.

Moreover, according to the system representative condition, $A \oplus B$ has further irreducible **correlational** properties. For each angle ϕ and the corresponding angles ψ, χ, there is a correlational property of $A \oplus B$ which corresponds to the subspace $[\psi_+^A \otimes \phi_+^B] \oplus [\chi_+^A \otimes \phi_-^B]$: $A \oplus B$ has this property just in case its system representative is included in this subspace, and this is true when the system representative of $A \oplus B$ is spanned by $\psi^{A \oplus B}$. Each of these correlational properties is well defined, and for each there exists an interaction with $A \oplus B$ which would be suitable to serve as an observation of that property. It is natural to express such a correlational property as follows:

$$((S_\psi^A(\text{up}) \wedge S_\phi^B(\text{up})) \vee (S_\chi^A(\text{up}) \wedge S_\phi^B(\text{down}))).$$

But this has the disadvantage of falsely suggesting that $A \oplus B$ possesses this property if and only if it possesses either $(S_\psi^A(\text{up}) \wedge S_\phi^B(\text{up}))$ or $(S_\chi^A(\text{up}) \wedge S_\phi^B(\text{down}))$. For a typical ϕ (such as $\phi = \pi/4$), $A \oplus B$ possesses neither of these latter composite properties, because of the composite exclusion condition. Only for special values of ϕ (such as $\phi = \pi/2$) does $A \oplus B$ have not only the correlational property $((S_\psi^A(\text{up}) \wedge S_\phi^B(\text{up})) \vee (S_\chi^A(\text{up}) \wedge S_\phi^B(\text{down})))$, but also one or other "disjunct" of this property, $(S_\psi^A(\text{up}) \wedge S_\phi^B(\text{up}))$ or $(S_\chi^A(\text{up}) \wedge S_\phi^B(\text{down}))$.

Consider now the effect of the M_B measurement. For simplicity, this will be assumed initially to consist of a minimally disturbing interaction which couples system B to another "apparatus" system M_B and induces the following evolution in the joint state of $B \oplus M_B$

$$\phi_\pm^B \otimes \mu_0^B \to \phi_\pm^B \otimes \mu_\pm^B,$$

121

where μ_0 is the "ready-to-measure" state of M_B, and $\mu_\pm^B$ are eigenstates of the operator corresponding to the recording quantity on system M_B. The effect of the M_B measurement on the total state is then to induce the following evolution

$$\psi^{A\oplus B} = N_+(\psi_+^A \otimes \phi_+^B) + N_-(\chi_+^A \otimes \phi_-^B),$$

$$\psi^{A\oplus B} \otimes \mu_0^B \rightarrow N_+(\psi_+^A \otimes \phi_+^B \otimes \mu_+^B) + $$
$$N_-(\chi_+^A \otimes \phi_-^B \otimes \mu_-^B).$$

Provided that $|N_+|^2 \neq |N_-|^2$, the system representative of M_B is now either $[\mu_+^B]$ or $[\mu_-^B]$, according to whether that of $A\oplus B$ is $[\psi_+^A \otimes \phi_+^B]$ or $[\chi_+^A \otimes \phi_-^B]$. For the typical case, $(\psi_+^A, \chi_+^A) \neq 0$, and so the system representative of A is neither $[\psi_+^A]$ nor $[\chi_+^A]$. Indeed, since the M_B interaction does not couple M_B to A, the dynamical postulates of §2.3 imply that the system representative of A is unaffected by the M_B interaction. This may be verified by reexpressing $\psi^{A\oplus B\oplus M_B}$ as follows:

$$\psi^{A\oplus B\oplus M_B} = \cos\theta \cdot \zeta_+^A \otimes \left(\sin\phi(\phi_+^B \otimes \mu_+^B) + \cos\phi(\phi_-^B \otimes \mu_-^B) \right)$$
$$- \sin\theta \cdot \zeta_-^A \otimes \left(\cos\phi(\phi_+^B \otimes \mu_+^B) - \sin\phi(\phi_-^B \otimes \mu_-^B) \right).$$

Thus, the system representative of A remains either $[\zeta_+^A]$ or $[\zeta_-^A]$ according to whether that of $B\oplus M_B$ is

$$[\sin\phi(\phi_+^B \otimes \mu_+^B) + \cos\phi(\phi_-^B \otimes \mu_-^B)]$$
$$\text{or } [\cos\phi(\phi_+^B \otimes \mu_+^B) - \sin\phi(\phi_-^B \otimes \mu_-^B)].$$

Consequently, the properties of (A,B) are as follows:

$$(S_z^A(\text{up}), S_\phi^B(\text{up})) \text{ or } (S_z^A(\text{up}), S_\phi^B(\text{down}))$$
$$\text{or } (S_z^A(\text{down}), S_\phi^B(\text{up})) \text{ or } (S_z^A(\text{down}), S_\phi^B(\text{down})),$$

and $A\oplus B$ has one of the four corresponding reducible properties, for example, $(S_z^A(\text{up}) \wedge S_\phi^B(\text{up}))$. In addition, $A\oplus B$ retains the correlational property

122

$$(\, (S_\psi^A(\text{up}) \wedge S_\phi^B(\text{up}) \,) \vee (S_\chi^A(\text{up}) \wedge S_\phi^B(\text{down}) \,) \,),$$

by virtue of the weakening condition: But it still does not possess either "disjunct" of this correlational property, because of the composite exclusion condition. Note also that the M_B interaction has destroyed many other correlations between A and B. Prior to the M_B interaction, $A \oplus B$ had a property of the form $(\, (S_{\psi'}^A(\text{up}) \wedge S_{\phi'}^B(\text{up}) \,) \vee (S_{\chi'}^A(\text{up}) \wedge S_{\phi'}^B(\text{down}) \,) \,)$ for *arbitrary* ϕ': After the M_B interaction it retains only that property with $\phi' = \phi$.

At this point it is interesting to note that although one may legitimately ascribe a quantum state to A following the M_B interaction, this is *not* identical to its system representative. As will shortly be demonstrated, when the M_B interaction leaves B with $S_\phi^B(\text{up})$, the subsequent M_A interaction will certainly leave A with $S_\psi^A(\text{up})$. One can bring this certainty within the framework of the Born rules by ascribing quantum state vector ψ_+^A to A following an M_B interaction with result $S_\phi^B(\text{up})$. But the system representative of A following the M_B interaction remains either $[\zeta_+^A]$ or $[\zeta_-^A]$, and *not* $[\psi_+^A]$.

Finally, consider the M_A interaction. This will be assumed to proceed according to an interaction coupling A to M_A as follows:

$$\psi_\pm^A \otimes \mu_0^A \rightarrow \psi_\pm^A \otimes \mu_\pm^A.$$

Now, $\zeta_+ = \cos\psi \cdot \psi_+ - \sin\psi \cdot \psi_-$, $\zeta_- = \sin\psi \cdot \psi_+ + \cos\psi \cdot \psi_-$. Hence

$$\chi_+ = \cos\chi \cdot \zeta_+ + \sin\chi \cdot \zeta_-$$

$$= \cos\chi\cos\psi \cdot \psi_+ - \cos\chi\sin\psi \cdot \psi_- + \sin\chi\sin\psi \cdot \psi_+ \\ + \sin\chi\cos\psi \cdot \psi_-$$

$$= \frac{1}{2N_+N_-} (\cos2\theta\sin2\phi \cdot \psi_+ + \sin2\theta \cdot \psi_-).$$

Thus,

$$\psi^{A \oplus B \oplus M_B} = N_+(\psi_+^A \otimes \phi_+^B \otimes \mu_+^B)$$

$$+ \frac{1}{2N_+} \left[\begin{array}{l} \cos2\theta\sin2\phi \cdot (\psi_+^A \otimes \phi_-^B \otimes \mu_-^B) \\ + \sin2\theta \cdot (\psi_-^A \otimes \phi_-^B \otimes \mu_-^B) \end{array} \right]$$

$$= \frac{1}{2N_+} \left[\begin{array}{l} (1 - \cos2\theta\cos2\phi) \cdot (\psi_+^A \otimes \phi_+^B \otimes \mu_+^B) \\ + \cos2\theta\sin2\phi \cdot \ \ (\psi_+^A \otimes \phi_-^B \otimes \mu_-^B) \\ + \sin2\theta \cdot \ \ \ \ \ \ \ \ \ (\psi_-^A \otimes \phi_-^B \otimes \mu_-^B) \end{array} \right].$$

Therefore, the effect of the M_A-measurement on the total state is to produce the evolution

$$\psi^{A \oplus B \oplus M_B} \otimes \mu_0^A \rightarrow \psi^{A \oplus M_A \oplus B \oplus M_B}$$

$$= \frac{1}{2N_+} \left[\begin{array}{l} (1 - \cos2\theta\cos2\phi) \cdot (\psi_+^A \otimes \mu_+^A \otimes \phi_+^B \otimes \mu_+^B) \\ + \cos2\theta\sin2\phi \cdot \ (\psi_+^A \otimes \mu_+^A \otimes \phi_-^B \otimes \mu_-^B) \\ + \sin2\theta \cdot \ \ \ \ \ \ \ \ (\psi_-^A \otimes \mu_-^A \otimes \phi_-^B \otimes \mu_-^B) \end{array} \right].$$

The system representative of A is now either $[\psi_+^A]$ or $[\psi_-^A]$, whereas that of B remains either $[\phi_+^B]$ or $[\phi_-^B]$. The system representative of $A \oplus B$ is now either $[\psi_+^A \otimes \phi_+^B]$, $[\psi_+^A \otimes \phi_-^B]$, or $[\psi_-^A \otimes \phi_-^B]$. Consequently, $(A, B, A \otimes B)$ have one of the following triples of properties:

$$(S_\psi^A(\text{up}), \ S_\phi^B(\text{up}), \ (S_\psi^A(\text{up}) \wedge S_\phi^B(\text{up})) \ ), \ \text{or}$$

$$(S_\psi^A(\text{up}), \ S_\phi^B(\text{down}), \ (S_\psi^A(\text{up}) \wedge S_\phi^B(\text{down})) \ ), \ \text{or}$$

$$(S_\psi^A(\text{down}), \ S_\phi^B(\text{down}), \ (S_\psi^A(\text{down}) \wedge S_\phi^B(\text{down})) \ ).$$

It follows that if the result of the M_B, M_A interactions is to leave B with $S_\phi^B(\text{up})$, then A will certainly be left with $S_\psi^A(\text{up})$. Note that though $A \oplus B$ does not possess $(S_x^A(\text{up}) \wedge S_\phi^B(\text{down}))$, it may now possess $(S_\psi^A(\text{up}) \wedge S_\phi^B(\text{up}))$. Indeed, it will possess this latter property just in case B now possesses $S_\phi^B(\text{up})$; and that in turn will be true if and only if B possessed $S_\phi^B(\text{up})$

already after the M_B interaction (since the M_A interaction did not affect the properties of B).

Not only will the M_B, M_A interactions leave A, B with correlated spin properties, but the systems M_A, M_B will also acquire correlated recording properties. Specifically, if M_B acquires $\mathcal{P}_+^{M_B}$, then M_A certainly acquires $\mathcal{P}_+^{M_A}$. In the case as described so far, it is these resultant properties of M_A, M_B which are directly observed, rather than those of A, B. The correlation established between the spin-components of A, B may be accurately inferred merely by noting the properties $\mathcal{P}_+^{M_A}$, $\mathcal{P}_+^{M_B}$. Indeed, if the M_A, M_B interactions were not minimally disturbing for their respective spins, then they would produce a correlation *only* between $\mathcal{P}_+^{M_A}$ and $\mathcal{P}_+^{M_B}$, leaving A, B without the definite, correlated spin properties these record. In the present interpretation, this would still be consistent with the M_A (M_B) interaction being appropriate to serve as an *observation* of the ψ-spin (ϕ-spin) of $A(B)$. Moreover, in a more realistic but complex case, M_A, M_B may each be just the first link in a chain of amplification devices, every other link of which interacts with the preceding link by a measurement-type interaction. Each chain continues at least until the recording properties for some link may be directly observed.

The case in which the M_A, M_B interactions, though SMIs, are not minimally disturbing for their respective spins is of some interest, and so I will treat it now in more detail. Suppose, then, that the M_B measurement proceeds in accordance with the interaction

$$\phi_\pm^B \otimes \mu_0^B \to \eta_\pm^B \otimes \mu_\pm^B,$$

where $\eta_\pm^B \neq \phi_\pm^B$, but we still have $(\eta_+^B, \eta_-^B) = 0$, $(\eta_+^B, \eta_+^B) = (\eta_-^B, \eta_-^B) = 1$, and it follows that the $\eta_\pm^B$ are in fact eigenvectors of spin-component of B along some direction (say, the η direction). The M_B measurement now induces the following evolution:

$$\psi^{A\oplus B} = N_+(\psi^A_+ \otimes \phi^B_+) + N_-(\chi^A_+ \otimes \phi^B_-),$$

$$\psi^{A\oplus B} \otimes \mu^B_0 \to N_+(\psi^A_+ \otimes \eta^B_+ \otimes \mu^B_+)$$
$$+ N_-(\chi^A_+ \otimes \eta^B_- \otimes \mu^B_-).$$

Provided that $|N_+|^2 \neq |N_-|^2$, the system representative of M_B is now either $[\mu^B_+]$ or $[\mu^B_-]$, according to whether that of $A\oplus B$ is $[\psi^A_+ \otimes \eta^B_+]$ or $[\chi^A_+ \otimes \eta^B_-]$. Since the M_B interaction does not couple M_B to A, the system representative of A is unaffected by the M_B interaction – it remains either $[\zeta^A_+]$ or $[\zeta^A_-]$. However, the system representatives of both B and $A\oplus B$ *are* now affected by the M_B interaction, so that B no longer has the property $S^B_\phi(\text{up})$ [or $S^B_\phi(\text{down})$], and $A\oplus B$ no longer has the correlational property $(\,(S^A_\psi(\text{up}) \wedge S^B_\phi(\text{up})\,) \vee (S^A_\chi(\text{up}) \wedge S^B_\phi(\text{down})\,)\,)$. But note that $A\oplus B$ has now acquired the correlational property $(\,(S^A_\psi(\text{up}) \wedge S^B_\eta(\text{up})\,) \vee (S^A_\chi(\text{up}) \wedge S^B_\eta(\text{down})\,)\,)$ and $A\oplus M_B$ has now acquired the correlational property $(\,(S^A_\psi(\text{up}) \wedge \mathscr{P}^{M_B}_+) \vee (S^A_\chi(\text{up}) \wedge \mathscr{P}^{M_B}_-)$. It is these latter correlational properties which now become fundamental to the explanation of the result of the M_A measurement.

The M_A interaction is now assumed to proceed as follows:

$$\psi^A_\pm \otimes \mu^A_0 \to \eta^A_\pm \otimes \mu^A_\pm.$$

This time the effect of the M_A measurement on the total state is to produce the transition

$$\psi^{A\oplus B\oplus M_B} \otimes \mu^A_0 \to \psi^{A\oplus M_A \oplus B\oplus M_B}$$

$$= \frac{1}{2N_+} \left[\begin{array}{l} (1-\cos2\theta\cos2\phi)\cdot(\eta^A_+ \otimes \mu^A_+ \otimes \eta^B_+ \otimes \mu^B_+) \\ + \cos2\theta\sin2\phi \cdot(\eta^A_+ \otimes \mu^A_+ \otimes \eta^B_- \otimes \mu^B_-) \\ + \sin2\theta \quad\cdot(\eta^A_- \otimes \mu^A_- \otimes \eta^B_- \otimes \mu^B_-) \end{array} \right].$$

The system representative of M_A is now either $[\mu^A_+]$ [μ^A_-], whereas that of M_B remains either $[\mu^B_+]$ or $[\mu^B_-]$. ~ystem representative of $M_A \oplus M_B$ is now either ~ $[\mu^A_+ \otimes \mu^B_-]$, or $[\mu^A_- \otimes \mu^B_-]$. Consequently,

$(M_A, M_B, M_A \oplus M_B)$ have one of the following triples of properties:

$$(\mathcal{P}^{M_A}_+, \ \mathcal{P}^{M_B}_+, \ \mathcal{P}^{M_A}_+ \wedge \mathcal{P}^{M_B}_+), \ \text{or}$$

$$(\mathcal{P}^{M_A}_+, \ \mathcal{P}^{M_B}_-, \ \mathcal{P}^{M_A}_+ \wedge \mathcal{P}^{M_B}_-), \ \text{or}$$

$$(\mathcal{P}^{M_A}_-, \ \mathcal{P}^{M_B}_-, \ \mathcal{P}^{M_A}_- \wedge \mathcal{P}^{M_B}_-).$$

It follows that if the result of the M_B, M_A interactions is to leave M_B with $\mathcal{P}^{M_B}_+$, then M_A will certainly be left with $\mathcal{P}^{M_A}_+$. Therefore, the outcomes of the M_A, M_B interactions recorded by M_A, M_B will display the same conditional certainty even when these interactions are *not* minimally disturbing, and so will neither reveal nor create the corresponding spin properties of A and B, respectively.

The foregoing discussion assumed that $|N_+|^2 \neq |N_-|^2$. What happens if this condition fails? The case $|N_+|^2 = |N_-|^2$ is just the kind of degenerate case treated in the discussion of measurement in §3.2. As explained there, interactions with the environment will lift the degeneracy and secure a definite outcome of the M_B interaction. But this now raises the worry that such interactions might also disturb the correlational properties on which the account of the conditional certainties in the outcomes of the M_A, M_B measurements depends. However, this worry is groundless. For the interaction couples only M_B (and not A or B) to the environment, and it must do so in a way which preserves any value of the recording quantity on M_B, if this is to provide a record of an outcome of the M_B measurement. The stability condition of §2.3 then guarantees that $A \oplus M_B$ continues to have the correlational property $(\ (S^A_\psi(\text{up}) \wedge \mathcal{P}^{M_B}_+) \vee (S^A_x(\text{up}) \wedge \mathcal{P}^{M_B}_-)\ )$, despite the interaction with the environment which lifts the degeneracy due to the fact that $|N_+|^2 = |N_-|^2$, and produces a definite outcome of the M_B interaction. And the free evolution condition of §2.3 guarantees that $A \oplus B$ continues to have $(\ (S^A_\psi(\text{up}) \wedge S^B_\eta(\text{up}))\ \vee (S^A_x(\text{up}) \wedge S^B_\eta(\text{down}))\ )$.

Now, $|N_+|^2 = |N_-|^2$ if either $\phi = \pi/4$, or $\theta = \pi/4$. The case

127

$\phi = \pi/4$ (introduced by Kochen) has been shown to introduce no difficulties for the account. But what about the original Bohm case, for which $\theta = \pi/4$? The only new feature introduced by this case is that here, prior to the M_B interaction, neither A nor B has any nontrivial spin property such as S_z^A(up) or S_z^B(down). This is because the initial system representative of $A \oplus B$ is a degenerate superposition, and so the system representative of each of A,B spans its entire spin space. Thus, whereas in the case $\theta \neq \pi/4$, A and B do initially have anticorrelated values of z-spin, in the case $\theta = \pi/4$, they have no such values. But it should not seem surprising that, when $\theta = \pi/4$, measurements of z-spin will produce perfectly anticorrelated outcomes even though neither A nor B had any definite z-spin prior to these measurements. For even in the $\theta \neq \pi/4$ case, it is only measurements of z-spin which reveal possessed values: Measurements of spin components along any other matched pair of directions (ϕ,ψ) will also reveal conditional certainties, even though these correspond to no preexisting values of ϕ-spin and ψ-spin.

Coupled spin-½ systems display correlations between the results of measurements on spin-components of the two systems besides those which correspond to conditional certainties or perfect anticorrelations. For an *arbitrary* pair of directions (ϕ_A,ϕ_B) there can be expected to be some correlation between the results of measurements of the component of spin of B along ϕ_B and the component of spin of A along ϕ_A. Indeed, these correlations are such that for certain states of the coupled systems and certain triples of directions (ϕ_1,ϕ_2,ϕ_3) they violate so-called Bell inequalities such as the following:

$$\text{prob}(A1;B3) + \text{prob}(A2;B3) + \text{prob}(A2;B1)$$
$$\leq \text{prob}(A2) + \text{prob}(B3), \qquad (4.1)$$

where, for example, prob($A1;B3$) is the joint probability of observing the spin components of A in the ϕ_1 direction and B in the ϕ_3 direction both to be positive, and prob($A2$) is the probability of observing the spin component of A in the ϕ_2 direction to be positive. For example, in the singlet state, when (ϕ_1, ϕ_2, ϕ_3) are coplanar and oriented at equal angles to one another, $\text{prob}(A1;B3) = \text{prob}(A2;B3) = \text{prob}(A2;B1) = 3/8$, $\text{prob}(A2) = \text{prob}(B3) = 1/2$, and the inequality (4.1) is violated.

How can one account for violations of Bell inequalities like (4.1) in the present interpretation of quantum mechanics? Note first that probabilities such as those just given follow directly from the Born rules as applied to the quantum spin state of the joint system $A \oplus B$. Insofar as quantum mechanics incorporates the Born rules, it is a straightforward consequence of quantum mechanics in the present interpretation that for certain states of $A \oplus B$, Bell inequalities such as (4.1) will be violated. But a more complete account of such violations may be provided by describing how a probability such as $\text{prob}(A2;B1)$ depends upon the dynamical states of the various systems involved in a joint measurement of the ϕ_2 spin-component of A and the ϕ_1 spin-component of B. Suppose, then, that the ϕ_1 spin-component of B is first measured on a system $A \oplus B$ in the singlet state. The M_B interaction produces the following evolution, as before:

$$\psi^{A \oplus B} = N_+(\psi_+^A \otimes \phi_+^B) + N_-(\chi_+^A \otimes \phi_-^B)$$

$$= (1/\sqrt{2}) \left((\phi_{1+}^A \otimes \phi_{1-}^B) - (\phi_{1-}^A \otimes \phi_{1+}^B) \right),$$

$$\psi^{A \oplus B} \otimes \mu_0^B \rightarrow$$

$$(1/\sqrt{2}) \left((\phi_{1+}^A \otimes \eta_{1-}^B \otimes \mu_-^B) - (\phi_{1-}^A \otimes \eta_{1+}^B \otimes \mu_+^B) \right).$$

As before, after interaction with the environment, M_B acquires the property $\mathcal{P}_+^{M_B}$ (with probability $1/2$), $A \oplus B$ definitely acquires the property $((S_\psi^A(\text{up}) \wedge S_\eta^B(\text{up})) \vee (S_\chi^A(\text{up}) \wedge S_\eta^B(\text{down})))$, and $A \oplus M_B$ acquires the correlational property $((S_\psi^A(\text{up}) \wedge \mathcal{P}_+^{M_B}) \vee (S_\chi^A(\text{up}) \wedge \mathcal{P}_-^{M_B}))$. These same correlational properties are now fundamental to the explanation of the probability of the A-measurement result, conditional on a positive B-measurement result.

The effect of the M_A interaction is now to produce the evolution

$$\psi^{A \oplus B \oplus M_B} \otimes \mu_0^A \rightarrow$$

$$\psi^{A \oplus M_A \oplus B \oplus M_B} = \begin{bmatrix} c_{++} \cdot (\eta_+^A \otimes \mu_+^A \otimes \eta_+^B \otimes \mu_+^B) \\ + \; c_{+-} \cdot (\eta_+^A \otimes \mu_+^A \otimes \eta_-^B \otimes \mu_-^B) \\ + \; c_{-+} \cdot (\eta_-^A \otimes \mu_-^A \otimes \eta_+^B \otimes \mu_+^B) \\ + \; c_{--} \cdot (\eta_-^A \otimes \mu_-^A \otimes \eta_-^B \otimes \mu_-^B) \end{bmatrix},$$

where $c_{++} = (1/\sqrt{2})(\sin\phi_1\cos\phi_2 - \cos\phi_2\sin\phi_2) = (1/\sqrt{2})\sin(\phi_1 - \phi_2)$ $(=\sqrt{3}/2\sqrt{2})$. It follows that the probability that M_A acquires $\mathcal{P}_+^{M_A}$ and that M_B acquires $\mathcal{P}_+^{M_B}$ is just $(1/2)\sin^2(\phi_1 - \phi_2)$ $(=3/8)$; and this is prob($A2;B1$). Hence, whereas the probability that M_A acquires $\mathcal{P}_+^{M_A}$ is $1/2$, the probability that M_A acquires $\mathcal{P}_+^{M_A}$ *given* that M_B acquires $\mathcal{P}_+^{M_B}$ is $3/4$. How does the probability that M_A acquires $\mathcal{P}_+^{M_A}$ depend in this way on the M_B result, even though the M_B interaction in no way affects the properties either of A or of M_A? The answer is that the M_B interaction *does* affect the compound systems $A \oplus B$ and $A \oplus M_B$ of which A is a component: $A \oplus B$ now acquires the correlational property $(\,(S_\psi^A(\text{up}) \wedge S_\eta^B(\text{up})\,) \vee (S_x^A(\text{up}) \wedge S_\eta^B(\text{down}))\,)\,)$, and $A \oplus M_B$ acquires the correlational property $(\,(S_\psi^A(\text{up}) \wedge \mathcal{P}_+^{M_B}) \vee (S_x^A(\text{up}) \wedge \mathcal{P}_-^{M_B})\,)$. Even though no dynamical property of A is affected by the M_B interaction, $A \oplus B$'s disposition to produce a positive outcome in the M_A interaction *is* affected by the M_B interaction. This is possible since $A \oplus B$'s disposition is grounded not only in the dynamical properties of A (and of M_A), but also in the dynamical properties of $A \oplus B$ and $A \oplus M_B$, including, in particular, the correlational properties just mentioned. Hence, these same correlational properties are grounds both for the conditional certainty of a positive outcome in a measurement of A's ψ-spin, and also for $A{\oplus}B$'s conditional probabilistic disposition to give a positive outcome in a measurement of spin on A in the ϕ_2 direction.

In the present interpretation, this concludes the basic account of the origin of EPR-type correlations. But this account may still seem unsatisfying. In what sense have these remarkable correlations now been *explained*? What microphysical processes give rise to them? If a naive realist interpretation had been possible, the origin of the correlations would have been at the source: The two component systems A,B would leave their common source with all their corresponding spin-components anticorrelated, and this initial (anti)correlation would subsequently give rise to the correlated results of the M_A, M_B measurements. The present interpretation rejects this view of the origins of the correlations. In the generic case, the only correlated properties with which A,B leave their common source are S_z^A, S_z^B; but these do *not* give rise to the correlated results of the M_A,M_B measurements. Indeed, in the original Bohm case $(\theta = \pi/4)$, A,B leave their common source with *no* nontrivial correlated properties. What account can now be offered? The correlations are so remarkable that some account seems called for. Recall that for arbitrary ϕ, if the result of the M_B measurement is that B has spin-up along the ϕ_+-axis, then it is certain that the result of the M_A measurement will be that A has spin-up along the ψ_+-axis; and this is so despite the fact that prior to the M_A measurement there is no property of A which grounds this certainty concerning its subsequent behavior in the M_A interaction.

In the present interpretation, this certainty concerning the subsequent behavior of A is grounded in irreducible correlational properties of compound systems of which A is a part. There are two such systems in this case, $A{\oplus}B$ and $A{\oplus}M_B$. Although I restrict attention to $A{\oplus}B$ in what follows, exactly parallel considerations also apply to $A{\oplus}M_B$. After

131

the M_B interaction, $A \oplus B$ has the correlational property ((S_ψ^A(up) $\wedge$ S_η^B(up)) $\vee$ (S_χ^A(up) $\wedge$ S_η^B(down))) as well as the composite property (I^A $\wedge$ S_η^B(up)) (where I^A is the trivial property that A always has).

The following reasoning (concerning the situation after the M_B interaction, but before the M_A interaction) is tempting. B has the property S_η^B(up), and lacks the property S_η^B(down). This is consistent with $A \oplus B$ having the correlational property ((S_ψ^A(up)$\wedge S_\eta^B$(up)) $\vee$ (S_χ^A(up)$\wedge S_\eta^B$(down))) only if $A \oplus B$ in fact has the first "disjunct" of this property, namely, (S_ψ^A(up)$\wedge S_\eta^B$(up)). But this in turn implies that A has S_ψ^A(up) even before the M_A interaction. If this conclusion were true, the M_B interaction would have changed the properties of A from having S_z^A(up) [or S_z^A(down)] to having S_ψ^A(up), despite the fact that the M_B interaction does not couple to A, and may occur when A is widely separated from B. In that case, the account would involve a nonlocal causal influence of just the sort postulated by a nonlocal hidden variable theory.

The conclusion of this reasoning is not true, however, since after the M_B interaction, but before the M_A interaction, $A \oplus B$ has neither "disjunct" of the property ((S_ψ^A(up) $\wedge$ S_η^B(up)) $\vee$ (S_χ^A(up) $\wedge$ S_η^B(down))), but only the "disjunctive" correlational property itself. But though it is fallacious to infer from $A \oplus B$'s possession of this correlational property, together with B's possession of S_η^B(up) rather than S_η^B(down), that A has S_ψ^A(up) even before the M_A interaction, it would be correct to conclude that A will be observed to have S_ψ^A(up) as a consequence of the M_A interaction. The certainty of this outcome of the M_A interaction is grounded, not in any prior property of A itself, but in prior properties of the compound system $A \oplus B$ of which A is a part [namely, the reducible property (I^A $\wedge$ S_η^B(up)), together with the correlational property ((S_ψ^A(up) $\wedge$ S_η^B(up)) $\vee$ (S_χ^A(up) $\wedge$ S_η^B(down)))].

The general pattern of explanation here may be illustrated by means of an analogy. If a rubber object is immersed for

132

some time in liquid nitrogen and then struck with a hammer, it breaks. Prior to immersion, it has a conditional disposition: If immersed in liquid nitrogen, it will become fragile. When the antecedent of this conditional is satisfied, the rubber object acquires the unconditional disposition of being fragile. The conditional disposition is grounded in some aspect of the molecular structure of the rubber, by virtue of which the rubber acquires the unconditional disposition of being fragile when it interacts in a certain way with another object (i.e., when it is immersed in liquid nitrogen). And the object's unconditional disposition (if it is acquired) is also grounded in some less permanent (temperature-dependent) aspect of its molecular structure. It may be that the rubber object is immersed if and only if some indeterministic process has one of two results. In that case, although the object has the conditional disposition whether immersed or not, it is a matter of chance whether or not it acquires the associated unconditional disposition. The explanation of the fact that the rubber object broke when struck is then the following: Prior to the outcome of the indeterministic process, the object had (by virtue of its molecular structure) the conditional disposition to become fragile if immersed in liquid nitrogen. As chance would have it, it was so immersed, and then became fragile due to changes in its molecular structure. It therefore broke when struck by the hammer.

In the case of the EPR-type correlations, the explanation of the ψ-spin-up outcome of the M_A interaction is similar. Prior to the M_B interaction, the $A \oplus B$ system had the conditional disposition to give result ψ-spin-up for the M_A interaction *if* the M_B interaction gave result φ-spin-up. It was a matter of pure chance that the M_B interaction did give φ-spin-up: But given that it did, the $A \oplus B$ system then acquired an unconditional disposition to give ψ-spin-up if subjected to an M_A interaction. The conditional disposition is grounded in the correlational property $((S_\psi^A(\text{up}) \wedge S_\eta^B(\text{up})) \vee (S_x^A(\text{up}) \wedge S_\eta^B(\text{down})))$; and the unconditional disposition is

133

grounded in this together with the further property (I^A $\wedge$ S_n^B(up)), acquired just in case the M_B interaction gives ϕ-spin-up.

The explanation of the correlations for unmatched (ϕ_A, ϕ_B) measurements, and hence of violations of the Bell inequalities, involves only one more idea: that of a probabilistic disposition. As explained in Chapter 1, an object has a probabilistic disposition to do α in circumstances C just in case, if it were put in circumstances C it would do α with probability p. Prior to the M_B measurement interaction, the $A \oplus B$ system had the conditional probabilistic disposition, with probability 3/4, to give result spin-up in an M_A interaction if subject to an M_B interaction which gives result spin-up. It was a matter of pure chance that the M_B interaction did give result spin-up: But given that it did, the $A \oplus B$ system then acquired an unconditional probabilistic disposition with probability 3/4 to give spin-up when subjected to the corresponding M_A interaction. These conditional and unconditional probabilistic dispositions are grounded in precisely the same properties of $A \oplus B$ (and $A \oplus M_B$) which were the grounds for the dispositions responsible for the conditional certainties for the matched pair of (ϕ, ψ) measurements.

One might object to this account of the origins of EPR-type correlations in the following way. Given that the outcome of the M_B interaction is ϕ-spin-up, the outcome of the M_A interaction is certainly ψ-spin-up. But the M_A interaction is appropriate for observing the ψ-spin of A: And an observation of a property can be certain to give a positive result if and only if the observed system in fact has the observed property. And yet the present interpretation denies that A has ψ-spin-up prior to the M_A interaction. Hence, the present interpretation is inadequate.

Note that since the conditional certainty about the outcome of the M_A interaction is interpretation-independent, this objection is equally powerful against any interpretation which denies that A has ψ-spin-up prior to the M_A interaction. But only on some variety of naive realist interpretation can one accept this prior existence of ψ-spin-up on A. The response to the objection is simply that in the case of quantum mechanical observation, it sometimes happens that an observation is certain to have a positive outcome, even when the observed system fails to have the observed property just prior to the observation-interaction. This was already noted as a possibility consistent with the quantum mechanical conception of measurement in the earlier discussion (Chapter 1). Here we see an important actual example of the phenomenon.

It has been assumed in the discussion so far that the M_B interaction occurs earlier than the M_A interaction. It is clear how that discussion could be revised to apply to the case in which it is the M_A interaction which comes first. But it may not be so clear how to handle the case in which the M_A, M_B interactions have no invariant time order, even though the predicted correlations are just the same in this case [and these predictions have now been verified by Aspect et al. (1982b)]. Now, one cannot expect a fully adequate treatment of this case within nonrelativistic quantum mechanics, however interpreted, since this theory presupposes an underlying Newtonian space-time structure which is incompatible with the existence of pairs of events with no invariant time order. But there is a natural extension of the present interpretation to this case which permits one to handle it in much the same way as the earlier cases.

The extension involves simply replacing discussion of moments of time by discussion of spacelike hypersurfaces. Thus, the subspace decomposition condition prescribes a relation between the system representatives of the universal system and those of its components on any spacelike hypersurface for which those components are not interacting. The condition may be applied given suitable assumptions concerning the space-time boundary of any such interactions. With this extension, one can now handle the case in which the M_A, M_B interactions have no invariant time order by describing the relevant dynamical properties of A, B, M_A, M_B, and their various compounds on each of a continuous family of nonintersecting hypersurfaces throughout the space-time region of the M_A, M_B interactions. Of course, one can only apply the subspace decomposition condition to yield such a description on a hypersurface for which the relevant systems are not interacting.

The complete description of the M_A, M_B interactions now consists of the conjunction of all of these partial descriptions. Among the partial descriptions will be some which are structurally identical to the earlier description of the case in which the M_B interaction occurs earlier, and others which are structurally identical to the case in which the M_A interaction occurs earlier. Thus, with the help of the above extension, the case where the M_A, M_B interactions have no invariant time order adds nothing fundamentally new.

5

Metaphysical aspects

As the previous chapter made clear, in the present interpretation quantum mechanics gives a rather specific account of EPR-type correlations and what underlies them. In the present chapter I argue that this account is genuinely explanatory (rather than merely an idle overlay on a simple description of the observed correlations), and that this constitutes a significant advantage of the present interpretation over certain rival interpretations, on which quantum mechanics is unable to offer an adequate explanation of the observed correlations. In order to make this argument it will prove necessary to examine the character of causal explanation in theoretical science, and in particular to develop a coherent metaphysics of a kind of holistic, nonseparable, causal explanation. I take the development of such a coherent metaphysics to constitute a significant task independent of the details of the present interpretation of quantum mechanics. Quantum mechanics is a mine of metaphysical insights, and I hope here at least to have exposed a rich seam.

The first task is to defend the adequacy of the present account of EPR-type correlations as at least a coherent description of what *may* underlie the observed correlations. I foresee three main challenges to its descriptive adequacy. The first is to claim that the account is incompatible with the relativistic requirement that there can be no direct causal connection between spacelike-separated events, and must therefore be rejected. A weaker version of this challenge is to claim that although the incompatibility just mentioned may not by itself entail rejection of the account, it does seriously undermine its claim to be genuinely explanatory. For it shows that there is no reason to prefer the present account to a naive realist or weak Copenhagen account, which would also be incompatible with this relativistic requirement. The second challenge is to claim that the account is inadequate since it is not clear that the processes it postulates can be given a relativistically invariant characterization. The third challenge is to claim that the account presupposes what is false, namely, that the subsystems A and B are spatially localized (at least approximately) throughout the experiments described (cf. Figure 1.1).

The second and third challenges are related and I shall respond to them first. In doing so it will be necessary to anticipate some of the ideas of Chapter 7 concerning the extension of the present interpretation to relativistic quantum theories. The statement of the subspace decomposition condition given in Chapter 2 presupposed an absolute simultaneity characteristic of a Newtonian space-time structure, since that condition constrains the dynamical state of a system at a *time* when it is not interacting. The natural generalization of this condition to a relativistic space-time structure will apply an analogous restriction to a spacelike hyperplane (or more general hypersurface) on which a given system is not subject to external interactions. When the generalized condition is applied to a system and all of its noninteracting subsystems on such a hyperplane, it will (in conjunction with the other conditions on dynamical property ascriptions) constrain the assignment of a dynamical state to the given system on each such spacelike hyperplane. Some of the properties in the dynamical state will be "spatial" (in a generalized sense): The system will have such a property if and only if it is confined to a given region of the spacelike hyperplane in question. It is such properties which (in the present interpretation) specify the extent to which a subsystem of the pairs considered in the previous chapter is spatially localized during the experiments described there. Without a detailed development of the present interpretation to accommodate a re-lativistic space-time structure, it is impossible to say just how localized these subsystems will be; and such a specification would in any case presuppose a statement of the Hamiltonian governing the motion of each subsystem (though perhaps this could be assumed to be just the free-particle Hamiltonian in each case). But there is certainly no reason to expect each subsystem to be localized in a manner which corresponds at all closely to that depicted in Figure 1.1: Each subsystem might be "smeared" over a much wider spatial region in the interval between the interactions, possibly even over an infinite spatial region. Fortunately, this in no way affects the adequacy of the present account. For that account proceeds by assigning properties (and most importantly, spin properties) to each subsystem on a series of spacelike hyperplanes, each of which intersects the causal future of the joint emission (or production) event e_s. No matter how small or how large a region of such a hyperplane a subsystem is confined to, the assignment to it of a spin property on that hyperplane will be perfectly well defined.

But even if localization is not a problem, one may still wonder whether the processes postulated by the present account can be given a relativistically invariant characterization. Before giving such a characterization, it is important to ask what relativistic invariance does and does not involve. One must first distinguish between the relativistic covariance of a formulation of a theory, and the relativistic invariance of a theory's description of the phenomena. Classical theories (such as electromagnetic theory, or the general relativistic treatment of gravitation) may be formulated as what Earman and Norton (1987) call "local space-time theories." Such formulations are not merely Lorentz covariant, but also generally covariant. Their general covariance is manifested by the fact that their basic quantities are four-dimensional geometric objects defined at each point within some open subset of the four-dimensional differentiable manifold which models space-time, and by the fact that their basic equations relate these quantities and their covariant derivatives at each point of the manifold where they are all defined. But the general covariance of such formulations does not guarantee that the theories themselves yield relativistically invariant descriptions of physical phenomena. The theory of Newtonian kinematics, for example, may be formulated as a generally covariant local space-time theory, even though this theory purports to describe physical phenomena in a way which is by no means relativistically invariant (since it appeals to an absolute time and a state of absolute rest). It is therefore necessary to introduce a further condition of Lorentz invariance which holds just in case, for any possible situation described by a theory, an active Lorentz transformation of that situation is also a possible situation described by the theory. Not every theory which may be formulated in a generally covariant manner meets this condition, and not every theory which meets this condition need be formulated in a generally covariant manner. A relativistic generalization of the present interactive interpretation of quantum mechanics therefore faces three questions: Does the theory describe physical phenomena in such a way as to meet the condition of Lorentz invariance? Can the theory be formulated as a local space-time theory? Can the theory be given a generally covariant formulation? In brief, I believe the answers to these questions are, respectively, "yes," "no," and "very probably, but such a formulation seems of little interest." I shall say something in defense of these answers in the context of the processes the theory postulates to occur in Aspect-type experiments.

Consider first how a local space-time theory specifies a process. One can give a geometric characterization of a process within such a theory by specifying the values of some quantity or quantities represented by a geometric object or objects defined at each point on a continuous nonspacelike curve, together with some covariant differential equations or other laws which these satisfy along the curve, and which

140

may be taken to govern the propagation of the relevant quantities along the curve. Or one might give such a specification on a whole family of such curves through a connected space-time region (corresponding, perhaps, to a congruence of continuous nonspacelike curves intersecting each of a continuous family of nonintersecting spacelike hypersurfaces). Note that if one does so specify the values of the relevant quantities on each hypersurface of one such family of spacelike hypersurfaces, the values on any other such family which "slices up" the region are then fixed automatically, since all these values derive ultimately from the values assigned at the space-time points which compose the hypersurfaces. One might further contemplate the possibility of "processes" characterizable by giving the values of some quantity or quantities at each point on a spacelike curve (or curves), together with some laws describing the "propagation" of the process along these spacelike curves. "Processes" of this second kind would be more problematic, and might be argued to be merely pseudo-processes, perhaps on the grounds that they lead to conflict with NIAD.

The process postulated by the present interactive interpretation of quantum mechanics to occur during Aspect-type experiments is specified very differently. Instead of representing a quantity by the assignment of a geometric object at each relevant space-time point, one now assigns dynamical properties associated with a quantity to a quantum system on each spacelike hyperplane (or, more generally, hypersurface) on which that system exists. A quantum system exists on a spacelike hyperplane if and only if it has some dynamical property on that hyperplane (including, in particular, the dynamical property of being located on that hyperplane, if not within some more restricted region of it). The spacelike hyperplanes are analogs to moments of time: Indeed, in Minkowski space-time, each hypersurface with Euclidean spatial geometry (each hyperplane) does correspond to a simultaneity of some frame. To specify a process within a connected

141

region of space-time, one specifies it on each continuous family of nonintersecting spacelike hyperplanes which "slices up" that region. To specify a process on a continuous family of spacelike hyperplanes, one specifies the relevant dynamical properties of some quantum system or systems on each hyperplane in the family, together with some laws which these properties satisfy, which may be taken to govern the propagation of the process through the region on this family of hyperplanes. Note that when that specification no longer derives ultimately from an assignment of values at space-time points, specifying a process on one family of spacelike hyperplanes does *not* automatically fix the assignment of dynamical properties on every other family of spacelike hyperplanes which "slices up" the region. A specification must now be given independently for each such family of spacelike hyperplanes, though the laws governing the relevant dynamical properties may make it possible to give this specification in quite general terms. I shall call a process **nonseparable** if it may be specified in the way just described but *not* by assigning a value to each relevant quantity at each space-time point within the region where the process occurs.

In the example under consideration here, the process involves properties associated with spin-components of the systems A, B, and $A \oplus B$, and may be *partially* specified in a space-time region including e_A, e_B, and e_S by giving the values of the relevant spin properties of these systems on each of an arbitrary continuous family of spacelike hyperplanes of one of two kinds. Define the **causal future** of a space-time region to consist of every space-time point which lies within or on the future light cone of some point in that region, though *not* itself lying within the region (and analogously for the causal past of a region). Every hyperplane of each of the two kinds in question intersects the causal future of e_S. The first kind of family contains hyperplanes which intersect the causal future of e_A while intersecting the causal past of e_B, but none which intersect the causal future of e_B while intersecting the causal past of e_A. The second kind of family contains hyperplanes which intersect the causal future of e_B while intersecting the causal past of e_A, but none which intersect the causal future of e_A while intersecting the causal past of e_B. It follows from (the natural relativistic generalization of) the stability condition of Chapter 2 that, given that there is no interaction with the spin properties of A, B, or $A \oplus B$ except, possibly, within e_S, e_A, or e_B, these spin properties remain unchanged over the hyperplanes corresponding to parameter values between e_A and e_B for any family of hyperplanes of either of these two kinds. This is the law which may be taken to govern the propagation of the process between e_A and e_B. It is to be supplemented by laws governing the overall transitions induced in the relevant spin-properties at e_A and e_B, which will be relativistic generalizations of those specified in the previous chapter. The *complete* specification of this process cannot be given in the absence of information concerning how the dynamical states of A, B, and $A \oplus B$ change on hyperplanes which intersect one or both of the regions e_A, e_B. But it will turn out that a more complete specification is not required to give an explanation of EPR-type correlations.

The process connecting e_S, e_A, and e_B in the present interpretation is not only nonseparable, but also **holistic**, in the following sense. There are stages of the process at which a system ($A \oplus B$) has dynamical properties which are not determined by dynamical properties of its parts (A and B); and the system's possession of these irreducible dynamical properties at this stage forms a part of the process, by virtue of the laws governing the process. Though these two features of the process – its nonseparability and its holism – are distinct, there is an interesting relation between them. Provided that the parts of a system are not spatially coincident throughout a process, the holism of the process implies its nonseparability. For if specifying the process involves ascribing an irreducible property to a system whose parts are not spatially coincident, then that process will be nonseparable, since these

143

specifications cannot be replaced by an ascription of properties at space-time locations on the world-lines of parts of the system. On the other hand, a nonseparable process need not be holistic. Even if every dynamical property of a system reduces to dynamical properties of its parts, their possession of those dynamical properties may itself fail to be spatially localized, in which case the specification of a dynamical property of a part will not reduce to ascriptions of properties at each space-time point on the world-line of that part.

Let me return now to the questions I raised earlier. Because the process postulated by the present interpretation to occur during Aspect-type experiments is nonseparable, quantum mechanics, in the present interpretation, cannot be formulated as a local space-time theory. For not only does the theory directly assign properties on extended regions of space-time (namely, spacelike hyperplanes or more general hypersurfaces), but it also fails to assign properties at the space-time points which compose these regions. Consequently, there is no way of formulating the theory as a local space-time theory, which would assign geometric objects at space-time points. It does not follow that the theory cannot be given a generally covariant formulation, however; merely that any such formulation must assign geometric objects not to space-time points, but to extended regions of space-time. Indeed, since these regions are specified in purely geometric terms (either as general spacelike hypersurfaces or as hypersurfaces of Euclidean spatial geometry), the specification of processes may itself be given in purely geometric terms. Since this is so, it should not prove impossible to give a generally covariant formulation of the theory that describes them.

The theory's description of these processes is Lorentz invariant, since applying an active Lorentz transformation to a possible process described by the theory yields another possible process described by the theory. This is obvious for the case of rotations and space-time translations. It also holds for Lorentz boosts, since a Lorentz boost applied to one of the processes depicted in Figure 1.1 yields another process of a type also depicted there. It follows from the nature of the laws governing these processes that such a boost will transform a type (a) process into another type (a) process, and a type (b) process into another type (b) process. This is because the laws governing the process are stated purely in terms of geometric relations in Minkowski space-time; and Lorentz boosts leave invariant both the light cone structure and the class of spacelike hyperplanes.

This concludes my defense of the claim that a relativistic generalization of the present interpretation of quantum mechanics would give a relativistically invariant characterization of the nonseparable process underlying the EPR-type correlations that occur in an Aspect-type experiment. The characterization could not be provided in greater detail without actually producing the generalization itself - a task which could not be undertaken here. Sketchy as it is, it should suffice to show that the second challenge can be met: A fully relativistically invariant characterization of the processes postulated by the present account may indeed be given.

There remains the first challenge: that the present account is incompatible with NIAD and must therefore be rejected. Recall the following statement of NIAD from Chapter 1:

NIAD: If two events e_a, e_b occur at spacelike separation from one another, then there can be no direct causal connection between them.

As it stands, this is not entirely clear: What counts as a causal connection, and what would make it direct? My strategy will be to treat the above statement as a schema, whose precise content may be clarified by considering arguments which may be offered for the principle, to see exactly what conclusion (if any) these justify.

As noted in Chapter 1, there is a tradition in the philosophy of space and time which considers causal principles to provide

145

a foundation for space-time theories, and NIAD, in particular, to be at least akin to a foundational principle of special relativity, such as the principle that light is a **first signal** – that, of all causal signals linking two distinct spatial locations, none can travel faster than the speed of light (in a vacuum). I take it that there are two main objections to this approach to special relativity. The first is that the central notions – of a causal signal, or of a (direct) causal connection – are unclear as they stand and at least in need of some explication. The second is that special relativity seems in need of no such foundation, logical or epistemological. The second objection is more fundamental, for it suggests that even if the key notions were successfully explicated the result would not be very interesting. But it is an objection that cannot be pressed here because it would lead too far from the central topics of this monograph.[1] I therefore fall back on the first objection: No clear sense has been given to the claim that NIAD is a foundational principle of relativity theory, and so there is no reason to believe that the present account is incompatible with any such foundational principle.

A second argument for NIAD proceeds by reductio ad absurdum. Starting from the negation of a version of NIAD one tries to show that quite innocent assumptions suffice to generate causal paradoxes, in which something happens if and only if it does not happen (e.g., a signal is sent by an observer or automatic signaling device at space-time point p if and only if it is *not* sent). Such arguments at least partially clarify the content of NIAD insofar as they spell out what is taken to follow from its violation. But there is still some residual unclarity, since it is not claimed that the causal paradoxes follow from that violation *alone*: Their generation still requires additional assumptions, however innocuous. Nevertheless, the generation of causal paradoxes requires that any violation of NIAD at least involves the possibility of some

---

[1]For a statement of this second objection I refer the reader to Friedman (1983).

kind of process propagating from one event to another event spacelike separated from it.

Now as the preceding discussion made clear, according to the present account, there *is* a process involving e_A and e_B whose description includes laws which may be taken to govern propagation between e_A and e_B. This suggests that this process could give rise to causal paradoxes. It is therefore important to see why it does not in fact do so.

Note, first, that in the case in which e_A and e_B are spacelike separated the process which connects them does so quite symmetrically. Whatever reason there is to accept that this process involves propagation from e_A to e_B, there is an equally powerful reason for accepting that it involves propagation from e_B to e_A; and the same holds with A, B systematically interchanged. In fact, there is some reason to accept that in this case the process involves propagation from e_A to e_B, and also from e_B to e_A, as we shall see. Now in order to generate a causal paradox, one would have to describe a situation in which an altered result of (say) e_B (or perhaps just an altered probability that e_B includes one result rather than the other) is propagated through the process in such a way that there is a corresponding alteration in a feature of e_A. The paradox would be generated by describing a situation in which this alteration is then propagated back through a process (perhaps involving some different pair A', B') to effect an alteration in some feature of whatever event in the causal past of e_B but not of e_A was responsible for the initially hypothesized alteration in e_B's result – and specifically an alteration just sufficient to undo this hypothesized alteration. But no such paradox can be generated, since whether the result of e_B is up or down is neither determined by nor probabilistically relevant to the features of any event in the causal past of e_B other than e_S (and events in its causal past): But all these events lie in the causal past of e_A. More succinctly (though less precisely), the result spin-down at e_B is an indeterministic event with the same probability as the result spin-up (conditional on e_S), and

147

this remains true irrespective of what else may or may not happen in the causal past of e_B. Consequently, one cannot use the result at e_B, together with the hypothesized process connecting this to the result at e_A, to describe a situation which, though at first sight may seem realizable, in fact turns out to be impossible since its description implies a contradiction. Hence, the process postulated by the present account does not lead to causal paradoxes.

Another way to see that this is so is to notice that if λ is restricted to a specification of the dynamical states of all systems wholly contained within the causal past of e_B, then the probability of a given result of e_B, given λ, depends only on the outcome of e_A, and not on the setting of M_A. It is the condition that Shimony (1986) calls outcome independence which fails for the present interpretation of quantum mechanics, whereas the condition he calls parameter independence holds. Though there has recently been some controversy over Jarrett's (1984) argument that a failure of parameter independence would entail a paradoxical violation of NIAD (see, e.g., Stairs, 1988), there has been widespread agreement that a violation of outcome independence would not. Anyone who shares in this agreement should also agree that the process postulated by the present account does not lead to causal paradoxes.

Nevertheless, in the case in which e_A and e_B are spacelike separated, the account does postulate a process linking spacelike separated events, and so it may be objected that it constitutes no improvement on other accounts which may be offered by more familiar approaches to quantum mechanics, such as naive realism or the weak Copenhagen view, which also postulate such a process. But the objection fails, because these accounts lead to additional serious problems of their own to which the present account is not subject.

Consider first the weak Copenhagen view. In this view, the quantum state of $A \oplus B$ also represents its dynamical state. The first spin measurement, either that on e_A or that on e_B, projects the state of $A \oplus B$ onto just one of the two components of the original superposition. If e_B occurs first and gives the result spin-down, then after e_B the state of $A \oplus B$ is a joint eigenstate of spin-down on B, and spin-up on A. Consequently, A then acquires spin-up, and this is indeed the result at e_A. The process of reduction then links spacelike separated events which occur simultaneously with the first measurement.

The difficulty with this account is not simply that it involves a process that links spacelike separated events. It is rather that this process cannot be given a relativistically invariant characterization. If e_A and e_B are spacelike separated, neither invariantly precedes the other: Which event then provokes the reduction of $A \oplus B$'s state? Even in a case in which e_A and e_B are timelike separated, with e_B (say) occurring invariantly earlier, in what region of A's "trajectory" does A acquire a definite spin property? These questions have no relativistically invariant answer. In order to give an answer one would have to single out some preferred state of motion corresponding to a frame whose coordinate time may be taken to define the true time-ordering of events such as e_A and e_B. Although it may not be inconsistent with the structure of relativistic space-time that there be such a frame, its existence would certainly violate the spirit of relativity. For it is normally taken to be Einstein's great achievement to have eliminated the need to assume the existence of any such privileged state of motion in formulating a fundamental physical theory. Moreover, the privileged frame in which reduction is instantaneous would be inaccessible to experiment, just like the hypothetical ether frame of Lorentz. On either realist or verificationist principles it counts as a serious epistemological defect in the weak Copenhagen interpretation if it requires the assumption of such a frame.

The significant difference between such an interpretation and the present account is this: Though both accounts involve a change in the properties of $A \oplus B$ immediately following (say) e_B, the weak Copenhagen view is forced to locate this change on some *particular* spacelike hyperplane through the "trajectories" of A and B, since it is accompanied (indeed, partly constituted) by a corresponding change in the properties of A. But in the present view no change in A accompanies the change in $A \oplus B$, and so the properties of $A \oplus B$ can consistently be supposed to change on *each* of a continuous infinity of spacelike hyperplanes which touch the future boundary of e_B. (In fact, this is merely an idealization of the present approach: Such hyperplanes actually delimit regions within which the corresponding change in $A \oplus B$'s properties will have occurred.)[2]

Consider now the naive realist approach, according to which each of e_A and e_B merely reveals the preexisting value of the spin-component measured, and the Born probabilities (for joint measurements as well as single measurements) are understood to correspond to the likely relative frequencies of these preexisting values. In order to reconcile this account to the violation of the Bell inequalities it is necessary to suppose either that after e_A the value of B's spin-component (at least in some directions) is altered, or else that after e_B the value of A's spin-component is altered. In the case in which e_A and e_B are spacelike separated, the interpretation then violates the principle of Einstein locality stated in Chapter 1: The real state of B *does* depend on what measurement (if any)

---

[2]In discussion, Gordon Fleming has pointed out that there is still a way open to the die-hard proponent of the weak Copenhagen interpretation who wishes to escape the preceding objection. He or she can admit that the dynamical properties of the separate systems A and B are themselves radically hyperplane-dependent, so that changes in the dynamical state of A as a consequence of the occurrence of e_B occur on hyperplanes which intersect only near e_B. In that case A would come to acquire a definite spin-component in very different regions of its trajectory, depending on which parallel family of hyperplanes one took to define successive moments of A's evolution. It would appear to be extremely difficult to reconcile this radical hyperplane-dependence of dynamical properties with the normal assumptions and experimental practices of physicists. It is a much easier task to accept such a radical hyperplane-dependence only for quantum states, interpreted in the way that they are in the present interactive interpretation as only weakly connected to dynamical states.

is carried out at A (or else, that of A depends on the B measurement). There is also a sense in which it violates NIAD, since it postulates a direct link between spacelike separated events: Either e_A produces a change in some property of B prior to e_B, or e_B produces a change in some property of A prior to e_A.

But would these violations be any more problematic than the process linking e_A and e_B postulated by the present account? Again, the answer is yes, since the links presupposed by the naive realist's account cannot be given a consistent relativistically invariant formulation. Consider first the case in which e_A occurs invariantly earlier than e_B. In what region of B's "trajectory" are its properties altered by the occurrence of e_A? The only relativistically invariant answer would appear to be "In that region where the boundary of e_A's causal future intersects B's trajectory." But this answer cannot be correct also for the case in which the two measurement events are spacelike separated. And indeed in that case, there is no relativistically invariant answer to the question "Which measurement event occurred earlier, thereby initiating an alteration in the properties of the other component of the pair?" The objection is substantially the same as that offered against the weak version of the Copenhagen interpretation. Moreover, even if one did accept a noninvariant answer to this last question, further unpalatable consequences would follow. In particular, there would be frames in which it is the *later* measurement event which initiates an alteration in the properties of the distant component at an *earlier* time.

I take it that these objections to accounts of the EPR-type correlations offered by the weak version of the Copenhagen view or by naive realism are sufficient to cast serious doubt on the adequacy of these accounts. Analogous objections against the present account of EPR-type correlations fail, principally because the irreducibility of certain of $A \oplus B$'s properties implies that changes in these properties do not involve changes in A or in B at uncertain and problematic space-time locations. But the present interpretation does not merely offer a consistent description of certain processes which may be involved in EPR-type correlations: This description permits one to *explain* these correlations, as I shall now argue.

An argument that a certain theory or account explains a certain class of phenomena may be divided into two stages. In the first stage, one simply assumes that the theory or account is true (or at least empirically adequate), and attempts to show that it does indeed bear the appropriate explanatory relation to the phenomena in question. But since no theory which is wholly unsupported by the evidence can be accepted as a satisfactory explanation of any phenomenon, it is further necessary to show that the evidence indeed provides substantial support for a theory before its explanation of the phenomena can be accepted. I shall therefore begin by assuming that the account of EPR-type correlations offered above is correct, in order to simplify the demonstration that it *would* explain these correlations *if* it were correct. Then I shall offer reasons for thinking that the evidence does in fact offer substantial support for this account.

Suppose that e_A lies wholly within the causal future of e_B. I claim that by the present account the result at e_B (together with the occurrence of e_S) causally explains the result at e_A. Consider first the case in which both spin-measurements are along the same direction. In this case, given an outcome of spin-down on B, an outcome of spin-up on A is certain (has probability 1). This outcome is certain insofar as it follows from the Born rules that whenever a singlet quantum state of a pair $A \oplus B$ is prepared by an event like e_S, the joint probability $P(A$ up and B down$)$ equals the single probability $P(B$ down$)$ $(= 1/2)$. Hence, the conditional probability

$$P(A \text{ up} \mid B \text{ down}) = P(A \text{ up } \& B \text{ down})/P(B \text{ down}) = 1.$$

This derivation from the Born rules might itself be claimed to be an explanation of the result at e_A. Certainly, if scientific explanation amounted simply to deduction from well-supported laws, this would be a natural claim to make. But there are at least two reasons for not accepting this derivation as an adequate explanation of the result at e_A. The first reason is that the laws which were appealed to (the Born rules) are themselves in need of explication within some interpretation of quantum mechanics, and no explanation in terms of them can be fully adequate before a satisfactory interpretation has been given. The second reason is that the derivation fails to answer important questions concerning how the result comes about at e_A, including the following one: How could features of e_S explain the perfect anticorrelations in the e_A, e_B results unless e_S gives rise to pairs $A \oplus B$ which themselves possess certain special properties? And yet no such properties are mentioned in the derivation. Following a spin-down outcome at e_B, what brings about the spin-up outcome at e_A? How can this outcome be uniquely determined by the spin-down outcome at e_B unless the latter induces some change in the world which gives rise to the former; and how could such a change give rise to this outcome unless there is some special mechanism by which it becomes linked to the outcome? Again, the derivation from the Born rules makes no mention of any such change or mechanism.

For both these reasons, derivation of a sentence expressing the spin-up outcome at e_A from the Born rules is, at best, an unsatisfactory and incomplete explanation of that result. But the present interpretation enables one to remedy its deficiencies and so to provide a more adequate explanation of the spin-up outcome at e_A. As shown in Chapter 3, the present interpretation offers a clear explication of the Born rules, thereby legitimating their employment in the derivation under consideration. More importantly, by specifying the character of the nonseparable process that links e_S, e_A, and e_B, the present account exhibits the "mechanism" by which the spin-up outcome at e_A comes about, and, consequently, offers a causal explanation of this outcome, albeit a causal explanation of a novel kind. Since I expect this last assertion to prove controversial, I shall now further explain and justify it.

If one reflects upon the character of a causal explanation of an event, two features stand out as of central importance. The first feature concerns the spatiotemporal continuity of the processes, appeal to which causally explains the event. To causally explain an event is to exhibit it as the outcome of a process (or processes) which continuously connect(s) that event to other events, objects, or processes in its spatiotemporal neighborhood. What makes the explanation causal is not merely the lawlike regularity instantiated by neighboring events or processes and the event to be explained, but also the whole interconnected set of lawlike regularities instantiated by the entire network of continuous processes which spatiotemporally connects them. This is not to say that all satisfactory causal explanations must *fully* describe all these regularities. One can give a satisfactory causal explanation of a worker's lung disease by noting that he worked in an asbestos mine for twenty years without adequate protection, and that the chance of developing lung disease is greatly increased by prolonged, unprotected exposure to asbestos. But this is satisfactory as a partial causal explanation only insofar as there is some more complete account to be found of the underlying processes connecting exposure to asbestos with development of lung disease.

But law-governed spatiotemporal continuity of connecting processes does not suffice for causal explanation. If it did, one could causally explain an event in terms of *later* events linked to it by such processes, or, equally, in terms of earlier events that are so linked. And, given suitable laws, one could causally explain an event in terms of events occurring, say, to the North of it at the same moment. For an event may be linked by law-governed spatiotemporally continuous processes to events which are later than it, or occurring simultaneously to the North of it, just as it may be linked by such

154

processes to earlier events. The second important character-istic, at least of causal explanations of events featuring some persistent object, is that they involve the actualization of a disposition of that object. An event featuring an object is causally explained by showing how continuous processes linking that object to earlier events in its neighborhood cul-minated in that object's having a disposition whose actuali-zation constituted the event to be explained. Now, the actualization of a disposition is a temporally asymmetric event, which occurs at a moment of time or over some in-terval of time: To say that a substance is soluble, for example, is to comment on its possible *future* behavior. And this is why a causal explanation of an event featuring a persisting object proceeds by linking that event to *earlier* events and processes involving that object rather than to later ones, or to simul-taneous events or processes occurring to the North of the event in question. It does not follow that an earlier event e featuring an object could never be explained in terms of later events (including f). But in order for this to be true, it would have to be possible to give content to the idea that, in addition to e preceding f with respect to the normal or global time order, there is some special or local time order with respect to which f precedes e, and with respect to which e's oc-currence constitutes the actualization of a disposition of the object linked by a continuous process to f (and other events). (If time travel were possible, but time travelers frequently experienced motion sickness, then this would be the pattern of a hypothetical causal explanation as to why they did!)

The present account, both of the perfect anticorrelations and of the spin-up outcome at e_A, given the spin-down outcome at e_B, has both of these features, and consequently provides a causal explanation of these phenomena. The non-separable process described above connects e_S and e_B to e_A. This process is spatiotemporally continuous in the following sense: The relevant spin properties of $A \oplus B$ remain constant over any continuous family of spacelike hyperplanes which intersect the causal future of e_S and the causal past of e_B, and also on any continuous family of spacelike hyperplanes which intersect the causal future of e_B and the causal past of e_A, and, finally, on any continuous family of spacelike hyperplanes which intersect the causal futures of both e_A and e_B. Moreover, any three such families which do not intersect may be "knitted together" into a continuous family linking e_S to e_A through e_B in such a way that the family may be parametrized by a continuous variable whose monotonic increase from e_S through e_A represents the passage of time relative to this family. Consequently, the outcome at e_A may be considered the actualization of a disposition of $A \oplus B$ over the interval of parameter values for this family corresponding to the event e_A. These causal explanations (of the perfect anticorrelations in the e_B and e_A outcomes, and of the e_A outcome given the e_B outcome) are not complete, insofar as they omit specifications of the linking processes on spacelike hyperplanes that intersect e_S, e_B, or e_A. But they are quite satisfactory as far as they go, and are certainly more complete and satisfactory than any mere derivation from the Born rules.

It is now time to discuss the explanation of correlations between the results at e_A and e_B other than the perfect anticorrelations, which occur only when the axes of the measuring devices are oriented along the same direction. If these axes are different, then the results will generally display statistical correlations, even though any combination of spin-up and spin-down outcomes at e_A, e_B may occur. The likely statistical correlations may once again be derived from the Born rules. The process postulated by the present account is just as before, up to the occurrence of e_A. The outcome at e_A still arises as the actualization of a disposition of $A \oplus B$: But this disposition is now a probabilistic disposition with probability p less than one. The process may be considered to propagate from e_S in such a way that $A \oplus B$ acquires a probabilistic disposition, with probability $p < 1$, to give a spin-up outcome at e_A. A spin-up outcome may or may not occur as

156

this disposition is actualized at e_A: The probability that it will occur is just p. The causal explanation *either* of a spin-up outcome *or* of a spin-down outcome at e_A consists in exhibiting that outcome as the actualization of $A \oplus B$'s probabilistic disposition, itself linked to e_S and e_B via the postulated nonseparable process. Note that a causal explanation may be offered of an event whose outcome was not determined. Note also that a causal explanation may be offered of an event which has a low probability (much less than 1/2) of occurring. In order to accept that the present account offers a causal explanation in such cases, it is necessary to reject both the view that a causal explanation may only be offered of an event whose occurrence was determined, and the (weaker) view that a causal explanation may only be offered of an event which was more likely to occur than not. Though both rejections seem entirely warranted (as reflection on this example may help to persuade the reader), I have nothing to add, by way of a general defense of these rejections, to Salmon (1984) and Lewis (1986).

So far, I have restricted attention to the case in which e_A lies wholly within the causal future of e_B. The case in which e_B lies wholly within the causal future of e_A is handled exactly symmetrically. But what about the case in which no part of e_A or e_B lies within the causal future of the other? How can the outcomes and their perfect anticorrelations now be causally explained? In fact, the causal explanations take the same form as for the case already considered. Despite the fact that there is no invariant time-ordering of e_A, e_B in this case, one may still causally explain the spin-up outcome at e_A in terms of e_S and the spin-down outcome at e_B by appeal to the nonseparable process linking these events. There is an infinite set of continuous families of spacelike hyperplanes each of which "slices up" the region in such a way that the time parameter relative to that family increases monotonically from e_S through e_B to e_A. The causal explanation of the spin-up outcome at e_A in terms of e_S and the spin-down outcome at e_B appeals to this aspect of the nonseparable process as indicating how e_S and e_B are linked to e_A through a spatio-temporally continuous process which gives rise to a disposition in $A \oplus B$ which is actualized at e_A. The causal explanation of the spin–down outcome at e_B in terms of e_S and the spin–up outcome at e_A appeals to the symmetric class of families of spacelike hyperplanes, each of whose relative time parameters increases monotonically from e_S through e_A to e_B.

It may seem strange that in this case one can both causally explain e_A in terms of e_B (and e_S), and also causally explain e_B in terms of e_A (and e_S), for causal explanations are typically asymmetric: If p causally explains q, then q does not causally explain p. Indeed, one might try to argue that *any* adequate causal explanation must be asymmetric, and that, therefore, the present account fails to yield adequate causal explanations for the case in which e_A and e_B lie wholly outside one another's causal futures. One such argument goes as follows. In giving a causal explanation, one explains one event in terms of another: This leaves room for an additional independent causal explanation of this second event, and so on. But if the relation of causal explanation between two events were symmetric, then the chain of explanation could close up in a tight circle, preventing any further explanations, and undercutting the value of even the initial explanation.

It is true that this openness is a nice feature of asymmetric causal explanations, but it seems to me that it is merely a prejudice to refuse to accept symmetric causal explanations as genuinely explanatory. Moreover, I suspect that this prejudice is closely connected to a prejudice in favor of determinism. If all events were determined by preceding events, then it is to such preceding events that one should look to explain the features of any given event. And the asymmetry of causal explanation then follows from the asymmetry of the *earlier than* relation. But in the present context it is particularly clear that belief in determinism *is* just a prejudice. If no earlier events determine that an event such as e_A will have the outcome it does have, then one cannot explain that outcome by appeal to such determining events. A symmetric causal explanation may then be the best explanation available, and certainly seems better than an appeal to mere randomness.

Another objection to symmetric causal explanations links them to causation. It goes as follows: e causally explains f only if e is the cause, or a cause, or a significant part of the cause, of f. However, causation is itself an asymmetric relation: If e and f are directly causally related, then either e causes f, or f causes e, but *not* both. Hence, causal explanation must also be an asymmetric relation.

I have deliberately postponed a discussion of the connection between causal explanation and causation until later. But even granted that it is as strong as the objection requires, why should causation itself be assumed to be an asymmetric relation? The philosophical literature is full of examples of pairs of events, each of which may plausibly be said to cause the other (consider, for example, the variation in the electric field strength and the variation in the magnetic field strength as an electromagnetic wave passes a point). If a pair of events are not simultaneous, then the earlier is typically (if not universally) the cause, and so examples of symmetric causal relations are hard to come by among such pairs. What if two events are simultaneous but not coincident? Then if NIAD holds, the pair cannot be directly causally related. But arguments for NIAD were examined earlier and found unconvincing, particularly in their application to the present case. Hence, there is no good argument, based on the premise that in such circumstances causation is always asymmetric, for the conclusion that when e_A and e_B lie wholly outside each other's causal futures the outcome at one cannot causally explain that at the other.

Nevertheless, there may still be those who object to the explanatory value of the present account on other grounds. One line of attack would be to claim that the nonseparable process postulated to link e_S, e_A, and e_B is merely a pseudoprocess, and any purported explanation that appeals to it is therefore merely a pseudoexplanation. The distinction between causal processes and pseudoprocesses goes back to Reichenbach (1956), but has recently been revived by Salmon (1984). In his book, Salmon draws this distinction in terms of the ability of a causal process (but not a pseudoprocess) to transmit a mark, and then develops an analysis of scientific explanation according to which a causal explanation appeals to or presupposes the existence of causal processes. For an approach like Salmon's, the key question to be answered concerning the nonseparable process postulated by the present account is whether or not it is capable of transmitting a mark. But this question proves to be a remarkably tricky one to answer; and the character of the difficulties involved is such as to warrant the conclusion that it is not necessary to give a positive answer to this question in order to show that the nonseparable process is indeed causal.

Note first that Salmon assumes that a marking interaction is an essentially local occurrence. Such a restriction seems clearly inappropriate for a nonseparable process, since a nonseparable process includes stages which occur not in a single local region of space, but over an extended region – and these stages need not be composed of parts which are so localized. It therefore seems natural to allow that a marking interaction for a nonseparable process may itself be a nonlocalized occurrence. The mark criterion for distinguishing causal processes from pseudoprocesses is explicitly modal: A causal process is one which is *capable* of transmitting a mark. How is one to decide whether a given process is capable of transmitting a mark? According to Salmon, this is a matter for experiment: If one randomly selects some instances of a type of process and alters a given stage in each, and if there is a regular difference between later stages of each process selected and the corresponding stages of instances not selected for alteration, then this establishes that processes of this type are capable of transmitting a mark. But such experiments can never definitively establish that a type of process is *not* capable of transmitting a mark – merely that no way has yet been found of using such a process to experimentally transmit a mark. Moreover, even a positive experimental result need not be accepted as definitive proof that a type of process is capable of transmitting a mark, since it may be that a type of process is so "fragile" that any alteration in an early stage prevents it from occurring at all, and gives rise instead to a quite different process.

It seems more satisfactory, therefore, not to appeal to experiment, but rather to the underlying laws obeyed by the type of process under investigation, taking experiment to be of value only insofar as it enables the experimenter to probe the nature of these laws. If the mark criterion is restated in terms of laws, a type of process may be said to meet that criterion if and only if the laws governing the process are such that for a range of variations in some features of an earlier stage or stages of the process, there will be some corresponding variation in the features of all later stages of the process.

But even when so restated, the criterion is still problematic. It is necessary to inquire into the source and character of the variations to which it refers. A "fragile" causal process will still fail to meet the restated criterion, provided that any variation in a feature of an earlier stage prevents the process from occurring at all. On the other hand, even when the restated criterion is met by a type of process, one may question whether that process is really capable of transmitting a mark, since, if a process is initiated by an indeterministic event, then even if variations in the features of this initiating event are propagated through later stages of the process, it may be incompatible with physical laws to suppose that these initial variations are controllable.

These considerations make it clear that there may be theoretical difficulties involved in supposing that a process may be marked at all, even though that process may naturally be described as propagating its structure from earlier to later stages. In such a case it seems correct to maintain that this process is causal, though problematic to say that it is capable of transmitting a mark. Consequently, though ability to transmit a mark may be regarded as excellent evidence that a process is causal, it should not be taken as a necessary condition for a process to be causal. What *is* necessary is that the process consist of spatiotemporally continuous sequences of stages, whose features vary in accordance with some law or laws, where each such sequence may be parametrized by a continuous, monotonically increasing variable which represents the temporal succession of the sequence.

Returning now to the nonseparable process linking e_S, e_A, and e_B in the present account of EPR-type correlations, what is one to say about whether that process is capable of transmitting a mark, and whether it is in fact a causal process rather than a pseudoprocess? I shall argue that although there are reasons for denying that it is capable of transmitting a mark, none of these is a reason to believe that it is a pseudoprocess rather than a causal process.

First, since e_A, e_B are indeterministic events (in the sense described above), stages of the process which overlap these cannot be marked. Nevertheless, variations in the relevant features of these stages are lawfully related to variations in features of other stages of the process parametrized by values corresponding to later times on some relevant family of hyperplanes. It is tempting to say that the process would be capable of transmitting a mark applied at e_A or e_B if only such marks could be applied: But this is problematic since the antecedent of the counterfactual is inconsistent with the physical laws governing the phenomena. Could intermediate nonseparable stages of the process be marked? There are reasons for supposing that they could not. It is certainly difficult to envisage any experimental method of marking these stages. Suppose, instead, one attempted to describe theoretical interactions capable of marking them. Such an interaction would presumably involve some perturbation in the spin properties of A, B, and $A \oplus B$ induced by an extra term in the Hamiltonian of the joint system $A \oplus B$.

Redhead (1987) has shown that the EPR-type correlations are not preserved under such perturbations. This constitutes a reason for supposing that the nonseparable process postulated as underlying these correlations is indeed so "fragile" that any attempt to mark its intermediate stages would prevent its occurrence entirely. But this reason may not be conclusive, since it cannot be required of a marking interaction that it leave the marked process entirely unaffected, but only that there be some relation between the mark and alterations in features of the later stages of the process. The question remains as to whether or not a perturbation in the features of an intermediate stage of the postulated process would give rise to a corresponding perturbation in later stages (relative to a suitable family of hypersurfaces). And, in fact, in the present account the answer to this question is positive: Any spin properties of $A \oplus B$ which emerged from such an interaction would remain stable for some time thereafter, in accordance with the stability condition of Chapter 2. However, it is still unclear whether the perturbation induced by an additional spin–dependent interaction term in the Hamiltonian of $A \oplus B$ would count as a *mark* in the postulated process, since such an interaction would induce *indeterministic* transitions in the dynamical state of $A \oplus B$, rather than some alteration uniquely determined by features of the interaction.

But though there is consequently some question as to whether or not the postulated nonseparable process is capable of transmitting a mark, there is no question that that process is a causal process, and not a pseudoprocess. For each relevant family of hyperplanes there are laws governing the continuous development of the dynamical state of $A \oplus B$ in the interval between e_A and e_B which link earlier stages of $A \oplus B$ to later stages (for that family); and these laws hold, independent of any other causal laws connecting properties of $A \oplus B$ or its parts to earlier events or processes. It is this last feature which distinguishes the postulated nonseparable process from a pseudoprocess. It is characteristic of a pseudoprocess that, although features of its stages may be related in a continuous and lawlike manner, this is so only because each stage is connected by independent causal processes to other events and processes that are not themselves parts of the pseudoprocess, so that the laws governing the pseudoprocess are conditional on the laws governing these independent causal processes – which is why any perturbation in a stage of a pseudoprocess fails to propagate to later stages of that process.

The process connecting e_S, e_A, and e_B in the present account is very different from a stereotypical causal process: It is a holistic, nonseparable process, which may be incapable of transmitting a mark. But I have argued that these are distinctive features of a novel type of *causal* process, and that they do not show that what is postulated by the account is merely a pseudoprocess. The postulated nonseparable process would indeed provide the basis for a causal explanation of EPR-type correlations, if it occurs. But is there any reason to believe that this process in fact occurs, or does its postulation yield a merely hypothetical explanation which there is no reason to take to be the correct explanation? Indeed, would it not be possible to "cook up" a merely hypothetical explanation of this type for *any* set of statistical correlations among pairs of distant events – in which case such an explanation seems too "cheap" to be taken seriously? Such skeptical doubts prompt an inquiry into the epistemological foundations of the present account, and, more generally, into the epistemology of irreducible property ascriptions.

To focus these doubts, consider how one might give an analogous explanation of the correlations between the results of measurements of angular momentum components on each of the two fragments of a satellite which was not rotating before it exploded. Assuming the system obeys the laws of classical mechanics, the results of measurements, on the two

164

fragments, of the angular momentum components along the same axis will be equal in magnitude but of opposite sign; and there will also be a strict correlation between the results of measurements on a pair along a different axis. One might postulate a holistic, nonseparable process to account for these results, which features irreducible correlational properties of the pairs such as *angular momentum component on* A *in direction* d *equals* S *if and only if angular momentum component on* A *in direction* d' *equals* T. Ascription of one such irreducible property for each pair of directions *d,d'* would permit one to describe a whole set of nonseparable processes linking explosion of the satellite to the results of measurements of angular momentum on its fragments. But it is quite clear that in this case there is no reason to believe that there are any such processes, and so there is no reason to take seriously an account which purports to explain the correlated results of the measurements in terms of such a process. Why should one take the above account of what underlies the EPR-type correlations any more seriously?

There are a number of important distinctions between the two cases which make all the difference to the justification for believing each of the accounts offered involving nonseparable processes. Perhaps the most salient difference is that, in the case of the exploding satellite, there is the following alternative explanation of the correlations offered by classical mechanics. The satellite as a whole, and also each of its fragments, all have a definite vector angular momentum at each moment, the vector sum of the angular momenta of the fragments is always equal to the angular momentum of the whole satellite (namely, zero), and this last quantity is conserved both during and after the explosion. Each measurement merely reveals the component of angular momentum of the measured fragment along the chosen direction, and this component is just the projection of the fragment's angular momentum vector along the chosen direction. This explanation makes implicit reference to a causal process linking the explosion to each of the measurement results, but this causal process is separable, and not holistic: It attributes a definite, conserved, angular momentum to each fragment as the two fragments separately spatially. Since this explanation is superior to an explanation in terms of holistic, nonseparable processes, it would be unwarranted to believe an explanation of the latter type in this case. There are three reasons why the separable explanation is superior. The first is that it is more economical: It requires the postulation of just three vector quantities in order to explain an infinite set of correlations. The second is that each of the postulated quantities may be manifested quite directly in suitable experiments. The third is that it is an explanation of an entirely familiar form, involving only conservation laws and local action. Therefore, in this case, although one can indeed "cook up" an explanation in terms of a holistic, nonseparable process, there is no reason to take that explanation seriously.

However, in the case of the EPR-type correlations, things are very different. In this case, not only is no explanation known involving only conservation laws and local action, but also, it seems to be a consequence of Bell's theorem that no such explanation is possible. Consequently, no explanation of a more familiar sort is even a rival candidate to that offered by the present account. Furthermore, the present account is quite economical in its postulation of irreducible properties: It does not postulate one such property for every correlation to be explained, but a *single* such property which figures in the explanation of every correlation, whether perfect or merely statistical. This fact may have been obscured by the way in which that property was referred to earlier (cf. Chapter 1). Even when d,d' are different directions, each of the following descriptions picks out the same correlational property: *Either A has spin-down in direction* d *and B has spin-up in direction* d, *or A has spin-up in direction* d *and B has spin-down in direction* d, and *Either A has spin-down in direction* d' *and B has spin-up in direction* d', *or A has spin-up in direction* d'

166

and B *has spin-down in direction* d'. The explanation of all the correlations thus involves only a single such property in this case, and is consequently very economical. Because of this economy, the present account of EPR-type correlations has a high degree of unifying power. The postulated irreducible properties do not merely mirror the observed correlations themselves, but unify them by exhibiting them all as consequences of a single underlying process that gives rise to them. It is not possible to "cook up" such a unified explanation of an arbitrary body of correlations involving pairs of distant events. I take it that the quality of an explanation which appeals to new theoretical structure varies roughly in proportion to the degree of unification it produces in the phenomena which prompted the postulation of this theoretical structure. If so, the uniformity of treatment of all the statistical regularities involved in the EPR-type experiments makes the present account into at least a passable explanation of them.

One might object that the unifying power of the present account is entirely derived from that already offered by quantum mechanics in more familiar interpretations: Surely, isn't it the ψ-function itself that really unifies the observed correlations insofar as they may all be derived from the Born rules applied to a single ψ-function (that of the singlet quantum state)? Given that all interpretations agree in attributing this unifying power to the ψ-function, how does the additional postulation of irreducible properties add to this unifying power?

Part of this objection may be admitted. It is true that the unifying power that results from postulation of irreducible properties is so intimately connected with that which flows from ascription of a single ψ-function that one cannot claim *additional* unifying power from this postulation. It is rather that in the present interpretation the unifying power inherent in the ascription of a single ψ-function is rendered comprehensible by explaining how the ψ-function of $A \oplus B$ is related

167

to the evolving dynamical properties of $A \oplus B$. Whereas other interpretations may appeal to the ψ-function as a unified repository of predictive information concerning the likely correlations in EPR-type experiments, none of the standard interpretations is able to connect the ψ-function to the dynamical properties of the quantum systems involved in such a way as to give an adequate causal explanation as to how these correlations come about.

In the strong version of the Copenhagen interpretation, there is no connection between a system's ψ-function at a time and the dynamical properties of any system at that time. In this interpretation, quantum mechanics has nothing to say about the dynamical properties of A, B, or $A \oplus B$ between e_S and e_A, e_B. Thus, if the strong version of the Copenhagen interpretation were accepted, quantum mechanics would be unable to give a causal explanation of the correlations in EPR-type experiments – the only "explanation" of these correlations forthcoming would consist in their derivation from the Born rules. But insofar as the Born rules exhaust the content of quantum state ascriptions in this interpretation, such an "explanation" would amount to no more than an appeal to "an empirical rule for predicting future correlations," to quote Bernard D'Espagnat (1979).[3] It could be accepted as

---

[3] The passage from which this quotation is taken is interesting. Here is a fuller quotation from D'Espagnat:

Whenever a consistent correlation between such events is said to be understood, or to have nothing mysterious about it, the explanation offered always cites come link of causality. Either one event causes the other or both events have a common cause. Until such a link has been discovered the mind cannot rest satisfied. Moreover it cannot do so even if empirical rules for predicting future correlations are already known. (p. 160)

The passage is interesting insofar as it recognizes only two possible kinds of causal explanation of EPR-type correlations: by appeal to a separable causal process linking (say) e_B to e_A, or by appeal to separable causal processes linking a common cause such e_S as both to e_A and to e_B. Although the present account of EPR-type correlations does indeed offer a causal explanation of them (or so I have argued), this explanation is of neither of these kinds: Instead it appeals to a nonseparable causal process linking e_A, e_B, and e_S.

adequate only by an instrumentalist for whom *all* scientific explanation amounts to no more than subsumption under a predictively useful generalization whose apparent commitment to deeper theoretical structure is to be disregarded. And even such an instrumentalist would owe us an account of the exact conditions under which quantum measurements lead to a definite outcome.

In the weak version of the Copenhagen interpretation there is indeed a close connection between a system's quantum state and its dynamical state: The former determines the latter. But (as was shown earlier in this chapter) this interpretation is unable to exploit this connection to give a coherent, relativistically invariant description of how the dynamical states of A, B, and $A \oplus B$ evolve between e_S and e_A, e_B; and, consequently, it is unable to offer a satisfactory causal explanation of EPR-type correlations.

For a naive realist interpretation, there is only a very weak connection between the quantum state of a system and its dynamical state. Consequently, the naive realist is unable to build a causal explanation of EPR-type correlations on top of the derivation of these correlations from the Born rules. Moreover, as was shown earlier in this chapter, the attempt to give an independent causal explanation which appeals to possessed and conserved spin-components of A and B fails not merely because it is forced to countenance causal processes linking spacelike separated events, but more importantly, because it is unable to give a relativistically invariant characterization of such processes.

By contrast, in the present interpretation, the connection between the ψ-function of $A \oplus B$ and its dynamical properties is closer (though still not as close as in the weak Copenhagen view), and this explains why appeal to the ψ-function leads to successful predictions concerning EPR-type correlations. But a direct appeal to the dynamical properties of $A \oplus B$ permits one to give a causal explanation of these correlations as resulting from the operation of a nonseparable causal process linking e_S, e_A, and e_B, which has a perfectly coherent and relativistically invariant characterization.

There remain two respects in which the present account of EPR-type correlations may be thought to be epistemologically defective. It may be claimed that the irreducible properties postulated by the account are not directly observable, and therefore their postulation is unwarranted. Or it may be claimed that although there is *some* warrant for postulating such properties, this is quite insufficient to justify accepting the present account of EPR-type correlations, since the nonseparable causal process it involves is so metaphysically strange and unlike any other known example of a causal process.

There is a sense in which it is true that the irreducible properties postulated by the present account are not directly observable, namely, that their observation requires instruments, and specifically measuring instruments. But in *this* sense most properties postulated in a basic theory like quantum mechanics are unobservable; and yet even contemporary empiricists admit that their postulation is unavoidable in theorizing, that theories that postulate them may be acceptable, and that such theories may be genuinely explanatory, whether or not there are any such properties.[4] Consequently, no specific reason has been offered as to why an explanation that postulates irreducible properties of compound systems is especially problematic. But if instruments are allowed, it is easy in the present interpretation to describe theoretical interactions with M-type instruments whose outcomes are governed by the irreducible properties of compound systems like $A \oplus B$. Such interactions could serve as measurements of irreducible properties, thus rendering them indirectly accessible to observation.

The objector will likely reject the significance of such ex-

[4] See, in particular, Van Fraassen (1980).

perimental results. After all, it might be claimed, it is only assuming the present interpretation that the results of such experiments establish the existence of irreducible properties: But these results may be explained quite simply without any such assumption. Hence, the appeal to results of such experiments as providing indirect evidence for the existence of irreducible properties is both circular and unnecessary.

Now, it has been a common practice of theoretical science at least since the time of Newton to appeal to experimental results as indirect observations of theoretical structures even when the significance of these results in doing so is established only by appeal to the theory which postulates them. Consider, for example, measurements of force based on Newton's second law of motion. This practice is not necessarily circular, insofar as it is possible to describe sets of results of such experiments which cannot be made to conform to the theory in question. The fact that such sets are not obtained provides support for the theory and justifies relying on it in interpreting the significance of these results. Moreover, support for the theory may be obtained from many independent sources, in each of which the theory is applied to an analysis of the observations in a different way. The concordance of the results of many and varied indirect observations of a theoretically postulated property justifies acceptance of each of them as indeed an observation of that property.

Of course, this assumes that no other theory is known which is equally capable of explaining all these experimental results without postulating the same theoretical structures. And a skeptic will likely claim that this is not so in the present case: After all, are there not many competing interpretations of quantum mechanics, in each of which it is possible to account for the results of those experiments which only the present interpretation takes to be indirect observations of irreducible properties of compound systems? But whereas the claim embodied in this rhetorical question is easy to make, it is not so easy to defend. The postulation of irreducible

171

properties of compound systems is not a mere detail of the present interpretation, invoked as an afterthought in order to make possible a hypothetical explanation of EPR-type correlations. Irreducible properties were first introduced into the present interpretation in quite a different context, in order to make possible a coherent treatment of measurement interactions. It is a significant advantage of the present interpretation over alternative attempted interpretations that it permits a coherent treatment of measurement interactions. Insofar as a coherent treatment of measurement interactions cannot be given in a more orthodox view such as the Copenhagen interpretation, it is not true that for such an interpretation it is possible to account for the results of M-type interactions that the present interpretation takes to be indirect observations of irreducible properties of compound systems. It is quite unclear whether there is *any* acceptable account of these results apart from that offered by quantum mechanics in the present interpretation.

In sum, the present account's postulation of irreducible properties of compound systems does not violate any reasonable empiricist strictures on the observability of theoretical structures: It is a consequence of quantum mechanics in the present interpretation that these properties are indirectly observable in certain M-type interactions; and it is at least doubtful whether quantum mechanics can give an adequate account of the results of these interactions in any other interpretation, since that would involve a solution (or dissolution) of the measurement problem.

One may still wonder whether there is sufficient warrant to accept the account of EPR-type correlations offered by the present interpretation. If this is the only available "causal explanation" of these correlations, would it not be better simply to accept that these correlations cannot be causally explained, but must be simply accepted as instances of natural regularities which may at least be subsumed under the Born rules of quantum mechanics? I hope that, after reading this far, many readers are convinced, as I am, that accepting the present account enhances their understanding of how EPR-type correlations come about. Such readers will agree with my claim that the present account offers an explanation of these correlations which is superior to any mere derivation from the Born rules. But however explanatory it may or may not be, the account offered by the present interpretation is not an optional part of this interpretation: Accepting the interpretation commits one to accepting this account. The important question, therefore, is whether or not to accept the overall interpretation that implies this account. And the answer to this question depends on more than conflicting intuitions on whether or not the present interpretation offers a causal explanation, or a good explanation, of EPR-type correlations. On the one hand, some alternative interpretations are unable even to give a coherent description of the processes involved in EPR-type experiments. On the other hand, its successful resolution of other conceptual problems in the foundations of quantum mechanics such as the measurement problem provides independent reasons for accepting the present interpretation, and therefore accepting its account of EPR-type correlations. It seems to me to be mere conservativeness and metaphysical timidity which prevents the acceptance of the present interpretation just because of the novel features of its account of EPR-type correlations. I shall conclude this chapter by exploring some consequences of this account for metaphysical issues concerning holism and the nature of causation. From a traditional philosophical perspective, these may well be the most interesting and significant aspects of the present interpretation. It is here that the metaphysician may expect to learn most from the interpretation, whether or not he or she comes to accept it as correct or superior to alternative interpretations. For metaphysics is concerned just as much with how the world might have been as with how it actually is.

Though claims are frequently made to the effect that a

certain theory, branch of science, or scientific approach is (or should be) holistic, the precise content of the claim and its metaphysical consequences are all too often quite unclear, so that rational assessment of the claim and its consequences is difficult if not impossible – though typically this does not prevent heated but confused debates! It is to be expected that philosophical readers will immediately become suspicious and even hostile when they learn that in the present interpretation quantum mechanics is committed to a kind of holism. It is therefore necessary to try to dispel such attitudes by a careful analysis of the holism involved in the present interpretation: Only after such reassurance will the philosophical reader be prepared to consider what metaphysical lessons may be learned from it.

All instances of holism within the present interpretation arise because a nonatomic quantum system may have irreducible dynamical properties – dynamical properties whose possession by that system does not supervene on the dynamical properties of its subsystems. If a system has such an irreducible property at a particular time, it may be said to be in a **holistic (dynamical) state** at that time. A stage of a process may be said to be a **holistic stage** if and only if it involves a system being in a holistic state. A **holistic process** is then a process that has at least one stage that is holistic. An explanation which makes reference (explicitly or implicitly) to a holistic process may be called a **holistic explanation**. All of these terms are consequently no more problematic than those used to define them, such as 'supervenience,' 'dynamical property,' 'time,' 'stage,' and 'involves.' Though it is not unreasonable to ask for further clarification of even these terms, neither is it unreasonable to claim that the above definitions represent philosophical progress in rendering talk of holism and holistic explanation relatively precise and unproblematic. Moreover, the above discussion of EPR-type correlations makes it clear that the present interpretation involves holistic explanations, processes, stages, and states of systems in these senses.

When it comes to events, things become at once more interesting and more problematic. The problems stem in large part from the absence of any clearly accepted philosophical treatment of events. Reflection on these problems may provide useful input into construction of such a theory. But these problems do not block a precise understanding of what it is for an event to be holistic. Indeed, there seem to be at least two rather different conditions for an event to count as holistic, but these are not clearly equivalent. Further metaphysical work is needed to sort out their relations and significance. On the one hand, an event may be said to be holistic$_1$ if it involves a change in an irreducible dynamical property of a system, or perhaps even the persistence of such a dynamical property on a system. On the other hand, an event may be said to be holistic$_2$ if it is composed of other events without being reducible to them; where an event e of type E is said to be reducible to another set of events $\{f_i\}$ $(i = 1, \ldots, n)$ of respective types $\{F_i\}$ just in case physical law does not allow for a set of events of types $\{F_i\}$ to occur and bear one another the same spatiotemporal relations as do the events $\{f_i\}$, even though no event of the type E then occurs. The intuition behind this second condition for an event to be holistic is as follows: If an event is holistic, then, although it may be composed of subevents, it is more than merely the sum of these subevents.

In Chapter 1, reference was made to "holistic happenings" involving the compound system $A \oplus B$ in EPR-type experiments. It is now possible to explicate this. If there is a change in some irreducible property of $A \oplus B$ because $A \oplus B$ either acquires or loses that property, then that change is an event involving $A \oplus B$, and, moreover, this is a holistic$_1$ event. If $A \oplus B$ continues to possess some irreducible property over a time interval, then its continued possession over some sufficiently short subinterval (or even at some moment during that interval) may or may not be considered to be an event. I introduce the term 'happening' to avoid deciding this question: The continued possession of an irreducible property, for a sufficiently short time, by a compound system, will be called a **happening**, as will be any event involving a change in an irreducible property of a compound system. Clearly, for the present interpretation, EPR-type experiments involve holistic$_1$ happenings, and at least some of these are holistic$_1$ events.

But do such experiments also involve holistic$_2$ events? This question cannot be resolved without agreement on a theory of events. The obvious candidate for the title of holistic$_2$ event is the compound event e_{AB}, composed of the subevents e_A and e_B. But in the absence of an agreed criterion for the existence and composition of events it is not clear that there is any such event, over and above e_A and e_B, themselves. *If* there is any such event as e_{AB}, and *if* it is composed of the parts e_A and e_B, then it is plausible to argue that it is indeed a holistic$_2$ event. It is consistent with physical law for events of the same types as e_A and e_B to occur unconnected by any holistic$_1$, nonseparable process involving $A \oplus B$: And if this happened, it is arguable that they would not compose an event of the same type as e_{AB}.

There is clearly a close connection between causal explanation and causation. Since the present interpretation introduces a novel kind of causal explanation, it is interesting to explore its metaphysical consequences for causation itself. Note, first, how the present account of the causal process linking e_S, e_A, and e_B in the EPR-type experiments satisfies the two metaphysical constraints on any causal relation between noncoincident events that were introduced in Chapter 1. The second of these constraints was that all pairs of noncoincident causally connected events must be connected by some causal process. In the present account, it is clearly consistent with this constraint to claim that e_B and e_A are causally connected events. But this leaves open the interesting question as to whether this implies that there is some asymmetric causal relation between these events. The second constraint was concerned explicitly with asymmetric causal relations. It was that there can be no asymmetric causal relation between noncoincident events unless these events are in some correspondingly asymmetric temporal relation.

In the case in which e_A and e_B are timelike separated, with e_A wholly within the causal future of e_B, e_B occurs absolutely earlier than e_A. Given that e_B and e_A are connected by a causal process, it would be in accordance with this constraint to say that e_B is a cause of e_A. Moreover, in this case e_B may be thought to effect a significant alteration in the process linking e_S to e_A, and to give rise through the operation of later stages of this process to a disposition of $A \oplus B$ which is actualized in e_A. If one focuses on its role in the causal history of e_A, one will find it quite natural, therefore, to call e_B a cause, or even the cause, of e_A, or at least of e_A's having the outcome it does have. But there is a different perspective on causation. In the present interpretation, even when e_B is absolutely earlier than e_A, it is impossible to use the process linking the two to affect the outcome at e_A (or even to affect its probability) by affecting e_B. There is no way of controlling the outcome at e_B, since it is an indeterministic event: But because the present interpretation satisfies the condition referred to earlier in this chapter as parameter independence, no change in what is measured at e_B has any effect on the probability of a given outcome at e_A. Therefore, if one focuses on the connection between causation and manipulatability, then one will be tempted to deny that e_B is a cause of e_A, despite the process linking e_B to e_A. Pending further metaphysical analysis it remains an open question in the present interpretation whether in this case e_B is a cause of e_A.

In the case where each of e_B and e_A lies wholly outside the causal future of the other, things are more interesting. In this case, even though neither event is absolutely earlier than the other, one could still regard e_B as a cause of e_A relative to any family of spacelike hypersurfaces for which e_B occurred earlier than e_A. By thus relativizing both temporal and causal priority to a family of hypersurfaces, one could still call one of e_A, e_B a cause of the other, while meeting the first constraint. Indeed, there is some temptation to do so if one focuses on that aspect of the connecting process which corresponds to one kind of family of spacelike hypersurfaces that compose it. For a family for which e_B precedes e_A, it is e_B's role in propagating a disposition to e_A which plays a part in the justification of the claim that the connecting process is causal. That role also provides some reason to count e_B as a cause of e_A, relative to that family. But again, if one focuses instead on the connection between causation and manipulatability, it is particularly clear in this case that one will count neither e_B nor e_A as a cause of the other, even relative to a family of hyperplanes.

Notice that e_S occurs invariantly earlier than both e_A and e_B; e_S is connected to each of e_A, e_B by a spatiotemporally continuous causal process which realizes dispositions of $A \oplus B$; and manipulation of e_S can affect the probability of a particular outcome of e_A or e_B (or indeed, whether either of these events occurs at all). Consequently, there seems to be every reason to say that e_S causes each of e_A, e_B. This is so despite the fact that e_S does not constitute a common cause of e_A, e_B which satisfies Reichenbach's (1956) principle of the common cause, since it does not screen off the outcome of e_B from that of e_A (or vice versa). I conclude [in agreement with Salmon (1984) and Cartwright (1988)] that Reichenbach's principle does not apply to *all* types of common cause.

There is an alternative way of attempting to connect the causal explanation of EPR-type correlations offered by the present account to causation. It is to claim that although in this account it is at least questionable whether either the outcome of e_B or that of e_A is a cause of the other, and perhaps even whether e_S is a cause of either outcome separately, nevertheless e_S does cause these outcomes to be correlated: It does so by causing the compound event e_{AB} to have (or to have with a certain probability) an outcome which consists of a correlated pair of individual outcomes.

The most serious objection to this approach would be to deny that there is any such event as e_{AB}. The validity of this objection cannot be determined without an agreed upon account of the existence conditions of events (or at least of *this* event). But if the existence of e_{AB} is granted, I see no objection to this approach, as long as it is viewed as supplementary, rather than complementary, to the present account. It is only if one who takes this approach further denies that there are any such events as e_A and e_B that it comes into conflict with the present account. But whether or not there is any such event as e_{AB}, it seems indisputable that there are such events as e_A, e_B. For example, e_A can be pointed to, referred to, and have its spatiotemporal bounds more or less precisely delineated (certainly precisely enough to be sure it does not overlap with e_B). More fundamentally, e_A has causes and effects which are all its own: Some of its causes are to be located in the sequence of events which resulted in the particular M_A setting selected, whereas its effects include the registration of the outcome of e_A. It is plausible to maintain that the existence of an event may be established by exhibiting events which are its causes and effects (indeed, it is popular to hold that the causes and effects of an event may be used to individuate it): If so, e_A's possession of its own proper causes and effects establishes the existence of e_A (and similarly for e_B). But if e_{AB} exists as well as e_A and e_B then it is difficult to avoid the conclusion that e_{AB} is composed of these latter events – in which case the present account is committed after all to the existence of holistic$_2$ events.

The last few pages have merely broached some of the fascinating metaphysical issues raised by quantum mechanics in the present interpretation. Their detailed exploration will more appropriately be pursued elsewhere.

179

6

Alternatives compared

After laying out the details of an interactive interpretation of (nonrelativistic) quantum mechanics, I shall now compare it to certain alternative approaches which have influenced its development. The main aim is to show how it differs from each of these, and to argue that these differences constitute significant improvements. But this explicit comparison may also foster a clearer understanding of the interactive interpretation itself, as well as showing how it is indebted to its predecessors.

§6.1 NAIVE REALISM

Let me begin with the naive realist interpretation. Recall that, according to the naive realist, the Born rules specify probabilities for possessed values of dynamical variables. This is intimately connected to the naive realist assumptions: that a quantum system always has a dynamical state specifying a precise real value for every dynamical variable (PV), and that this is the value which a successful measurement of that dynamical variable would reveal (FM).

Though denying all three of these assumptions, the present interactive interpretation nevertheless agrees with the naive realist account on certain significant points. As opposed to the Copenhagen interpretation, both interpretations agree that a quantum system always has a dynamical state which is not derived from any quantum state. Furthermore, both interpretations agree that "statements concerning measurements can occur only as special instances viz., parts, of physical description to which I cannot ascribe an exceptional position above the rest."[1] Thus, in neither the naive realist interpretation nor the present interactive interpretation are the fundamental probabilities in quantum mechanics probabilities *of measurement results*. Rather, in both interpretations, measurement involves just a special kind of physical interaction, subject to basically the same physical laws as other interactions.

But this agreement must itself be qualified. It is distinctive of the present interactive interpretation that the dynamical state of a compound quantum system may be *holistic*. And though, in the present interpretation, the fundamental probabilities in quantum mechanics do not concern measurements, nevertheless the (nonfundamental) Born rules do give (conditional) probabilities for the outcomes of a class of physical interactions which importantly includes interactions employed in performing quantum measurements.

These disagreements, as well as the qualified nature of the agreement, are critical to the reasons why the present interactive interpretation is to be preferred to a naive realist interpretation. Two such reasons have been noted already, in the Introduction and in Chapter 5. After briefly reviewing these here I shall add a third. Rightly or wrongly, it has often been assumed that any interpretation of quantum mechanics according to which the dynamical state of a quantum system is not derivable from any quantum state is ipso facto a hidden variable theory. And there is a long and confused history of arguments to the effect that no hidden variable theory can be compatible with those statistical predictions of quantum mechanics that follow from the Born rules. One can abstract from this history two powerful types of argument against a naive realist interpretation of quantum mechanics. But neither type has any force against the present interactive interpretation.

---

[1]This is quoted from Einstein's reply to his critics in Schilpp, ed. (1949), p. 674.

Theorems due to Gleason (1957) and Kochen and Specker (1967) have been used to argue convincingly that it is false that a quantum system always has a dynamical state which specifies a precise real value for each dynamical variable.[2] Such an argument presents formidable difficulties for any naive realist interpretation that incorporates assumption PV. But since the present interactive interpretation rejects PV, no such argument can be used to show that the present inter-pretation is incorrect. It is quite consistent with the results of Gleason and of Kochen and Specker to maintain (as does the present interpretation) that every quantum system has a dynamical state that is not derivable from its quantum state, but that this state does not specify a precise real value for each dynamical variable. There is no tension between these theorems and the structure of dynamical states spelled out in Chapter 2.

The second kind of "no-hidden-variable" result, due orig-inally to Bell, may be used to argue that any interpretation which ascribes a dynamical state to each quantum system must accept that the dynamical state of one system may be influenced instantaneously and nonlocally by an alteration in the dynamical state of another system which is physically and spatially separate from it. But whereas such an argument has force against a naive realist interpretation [cf. Chapter 5, as well as Healey (1979) and Redhead (1987)], it does not touch the present interactive interpretation for one basic reason. In the present interactive interpretation, an alteration in the dy-namical state of one component of a compound system may change the dynamical state of the compound without in any way affecting the dynamical state of a separate component. In a pair of compound systems of the type first considered by Einstein et al. (1935), this gives rise to a nonseparable

---

[2]Such arguments have been given by Healey (1979) and Redhead (1987), among others.

process connecting the two components without any non-locality or failure of relativistic invariance.

There is a third reason for preferring the present interactive interpretation to naive realism. The dynamical states postulated by the naive realist differ only in detail from the dynamical states of classical mechanics. But whereas classical mechanics also included (false) laws governing the evolution of its dynamical states, naive realism has nothing to say about the dynamical evolution of individual quantum systems. Of course, the naive realist may portray this silence as an advantage of his interpretation – that it honestly admits that quantum mechanics is incomplete, thereby encouraging attempts to complete it by further theorizing [cf. the final sentence of Einstein et al. (1935)]. But it is not at all clear how any naive realist completion of quantum mechanics could plausibly account for interference phenomena. Consider, for example, the notorious two-slit thought-experiment. If each electron passed through just one slit, then how (other than via physically implausible nonlocal action) could opening the other slit affect its dynamical behavior in such a way that it contributes to the observed interference pattern constituted by electrons detected in various regions of a screen placed behind the two slits? No such problem arises with the present interactive interpretation. Even though each individual electron in the two-slit experiment has a dynamical state, this does not imply that each electron passes through just one slit. Rather, each electron is acted on by a potential associated with both slits, and this induces indeterministic transitions in its dynamical state which in no way reduce to the transitions that would be induced by the potential associated with each single open slit. The passage of each electron through the apparatus is a nonseparable process, involving indeterministic transitions in a highly nonclassical dynamical state. Moreover, in the present interactive interpretation, the detection of each individual electron within a small region of the screen in no way implies that this electron was located within that region at or just prior to its detection, since the present interpretation does not incorporate the naive realist faithful measurement assumption FM. Moreover, this assumption is not simply dropped to avoid such unwanted consequences, but is rather replaced by a coherent dynamics of measurement-type interactions which provides a framework within which its partial failure may be naturally understood.

Although its treatment of the dynamics of an individual quantum system is in these respects different from and superior to that provided by naive realism, the present interactive interpretation shares with naive realism the view that quantum mechanics does not itself provide a *complete* dynamics of individual quantum systems. What is required to supplement the constraints on dynamical evolution inherent in quantum mechanics (under the present interpretation) is a detailed stochastic theory governing the indeterministic transitions among (highly nonclassical) dynamical states induced by interactions. Such a dynamical theory would be a far cry from a deterministic hidden variable theory!

§6.2 COPENHAGEN

Since the Copenhagen interpretation continues to be regarded as orthodoxy by many physicists, it is of particular interest to compare it to the present interactive interpretation. There are several important points on which the two interpretations agree. Both interpretations agree that for each quantum system there are always some dynamical variables which have no precise values, even though a well-conducted measurement of such a variable would yield a precise value: Moreover, both interpretations agree that conjugate quantities never have simultaneous precise values.[3] The interpretations

---

[3]In the present interpretation, no *atomic* system ever has such simultaneous precise

agree that measurement typically influences the dynamical properties of a system (including properties corresponding to possible values of the measured variable), rather than simply revealing properties the system possesses. And they also agree that the primary role of the mathematical representative of the quantum state is to generate statistical predictions of measurement results on elements of an ensemble in a well-defined experimental arrangement, and not to describe the dynamical properties of an individual quantum system.

But the differences between the interpretations are at least as important. One distinctive feature of the present interactive interpretation (as compared to the Copenhagen interpretation) is that quantum mechanics is *not* fundamentally about observations. Terms such as 'observation,' 'measurement,' and 'classical system' do not appear in a careful formulation of the basic principles of the theory. Rather, measurements are treated as elements of a class of physical interactions whose characteristic features may be specified quantum mechanically without any use of such terms. According to the present interactive interpretation, each quantum system always has a dynamical state (whether or not it is observed), and quantum mechanics is fundamentally concerned with probabilistic relations among such states. A further distinctive feature concerns the effects of a measurement interaction: A measurement typically does not create the measured value, and immediate repetition of a well-conducted measurement may or may not give the same result. A third distinctive feature concerns the quantum state. In the present interactive interpretation, the quantum state of a system describes some of its associated objective probabilistic dispositions, and not just our *knowledge* of these dispositions. Consequently, a change in the quantum state of a system upon measurement is not simply a consequence of updating our knowledge of that

values: Whether this is true for *all* systems remains an open question, whose answer requires further investigation of the constraints on dynamical states given in Chapter 2.

185

system in the light of further information concerning the result of the measurement. But neither is such a change the result of a sui generis physical process that occurs only on measurement ("collapse of the wave packet"). The probabilistic dispositions associated with a quantum system change upon measurement just because measurement involves a physical interaction, and any physical interaction tends to produce indeterministic transitions in the dynamical states of systems involved in that interaction.

On what grounds is the present interactive interpretation to be preferred to the Copenhagen interpretation? There are two related general reasons for such preference. The present interactive interpretation renders quantum mechanics a precisely formulated, explanatorily powerful theory; and it permits a precise and nonproblematic account of the role of measurement in quantum mechanics. I shall now explain and justify these claims.

Consider first what I called in the Introduction the *weak* Copenhagen interpretation, according to which for each system there is a quantum state which completely specifies its dynamical state according to the following prescription: A system has a dynamical property if and only if the associated quantum state assigns that property probability 1 (via the Born rules). At first sight this prescription may seem perfectly precise. But this apparent precision proves illusory, as can be shown by focusing on the association between a system and its quantum state. Note that this association does not take the form of a one-to-one correspondence. If, for example, a pair of spin-½ particles is in the singlet spin state, then neither particle has its own individual quantum state – rather, each is associated with the joint quantum state. Such failure of one-to-one correspondence between systems and quantum states will occur whenever the mathematical representative of such a joint quantum state cannot be expressed as a direct product of the representatives of its components. As Schrödinger (1935b) put it:

When two systems, of which we know the states by their respective representatives, enter into temporary physical interaction due to known forces between them, and when after a time of mutual influence the systems separate again, then they can no longer be described in the same way as before, viz. by endowing each of them with a representative of its own. I would call this not *one* but *the* characteristic of quantum mechanics. (p. 555)

Now if all interactions, no matter how weak or short-lived, produced such "entangled" quantum states, then (given the ubiquity of interactions) it would rapidly become impossible to associate any quantum state with a system except one describing the entire universe! Nevertheless, applications of quantum mechanics typically proceed by describing quite simple microscopic systems by their own quantum states. The apparent inconsistency to which these facts give rise may be called the **preparation problem**. How can this problem be solved under the weak Copenhagen interpretation?

One way to solve the preparation problem would be to show that it is an approximation, valid for most purposes, to describe restricted subsystems of the universe by their own simple quantum states. But no such demonstration is offered by proponents of the weak Copenhagen interpretation. Instead, the measurement problem provides a powerful reason why none can be forthcoming. For consider a measurement interaction between a quantum system α and an apparatus system σ. Suppose it were legitimate to describe each of σ, α by its own quantum state prior to their interaction. If the interaction is capable of effecting correlations between initial values of the measured variable on σ and final values of the "pointer reading" variable on α, then the postinteraction quantum state of $\sigma \oplus \alpha$ will typically be an "entangled" state, not expressible as a product of separate quantum states of σ and of α. Moreover, this joint quantum state of $\sigma \oplus \alpha$ will then be one in which, according to the weak Copenhagen interpretation, the "pointer reading" variable has no definite value on α – the pointer points nowhere at all.

187

This last problem can be solved in one stroke by denying that measurement is a process that conforms to the usual linear time-evolution law of quantum mechanics (the time-dependent Schrödinger equation). This is doubtlessly one reason why proponents of the weak Copenhagen interpretation since von Neumann have resorted to the projection postulate. Moreover, if measurement really reduced a superposition to a single component, one could justify the ascription of simple quantum states to restricted subsystems of the universe like atoms or molecules. For one could take such ascription to be legitimate following a *preparatory measurement*. This preparatory measurement would project the superposed object system + environment quantum state onto a single one of its components, namely, a product of two quantum states, the first of which could then be taken to be the quantum state of the individual object system after the measurement.

But attempting to solve the problems of state preparation and measurement by appeal to the projection postulate lead to what might be called the **characterization problem** – namely, the problem of specifying under precisely what circumstances the quantum state of a system evolves in accordance with the projection postulate rather than in accordance with the time-dependent Schrödinger equation. Without a satisfactory solution to the characterization problem, the weak Copenhagen interpretation remains intolerably imprecise. Attempts to solve the characterization problem by appeal to the intervention of consciousness, interaction with a classical system, or the irreversibility of the measurement interaction, have not succeeded.[4]

One might regard "consistency proofs" like that of von Neumann (1932), Chapter 6, as attempts to show that despite the characterization problem quantum mechanics is quite un-

---

[4]Although I believe that this claim is not particularly controversial, it would take me too far afield to justify it here.

ambiguous in its predictions, and that, therefore the weak Copenhagen interpretation does, in practice, yield a perfectly precise interpretation of quantum mechanics. The idea of such "proofs" is to show that the statistical predictions of quantum mechanics are the same no matter at what link in a chain of correlation interactions one supposes that a measurement has occurred, with the associated application of the projection postulate. Thus, consider a chain of ideal correlation interactions of the form

$$\chi_i^{n-1} \otimes \chi_0^n \rightarrow \chi_i^{n-1} \otimes \chi_i^n \quad (n = 1, 2, \ldots, N), \quad (6.1)$$

with $\chi^0 = \phi$ and $\chi^N = \psi$. Here it is assumed that the ϕ describe the object system O, and the ψ are states of a macroscopic system A which are directly perceptually distinguishable. Now each of the following sequences yields the same final compound quantum state, with probability $|c_k|^2$,

$$\Sigma c_i \phi_i \rightarrow \quad \phi_k \otimes \chi_k^1 \rightarrow \quad \phi_k \otimes \chi_k^1 \otimes \chi_k^2 \rightarrow \cdots$$
$$\rightarrow \phi_k \otimes \chi_k^1 \otimes \chi_k^2 \otimes \cdots \otimes \psi_k,$$

$$\Sigma c_i \phi_i \rightarrow \Sigma c_i (\phi_i \otimes \chi_i^1) \rightarrow \quad \phi_k \otimes \chi_k^1 \otimes \chi_k^2 \rightarrow \cdots$$
$$\rightarrow \phi_k \otimes \chi_k^1 \otimes \chi_k^2 \otimes \cdots \otimes \psi_k,$$

$$\Sigma c_i \phi_i \rightarrow \Sigma c_i (\phi_i \otimes \chi_i^1) \rightarrow \Sigma c_i (\phi_i \otimes \chi_i^1 \otimes \chi_i^2) \rightarrow \cdots$$
$$\rightarrow \phi_k \otimes \chi_k^1 \otimes \chi_k^2 \otimes \cdots \otimes \psi_k,$$

(and so on).

Hence, in such a chain, it does not matter which sequence actually occurs, insofar as the statistics for the perceptually distinguishable final states of A are the same for each sequence.

But this shows merely that in such a chain all these sequences are empirically indistinguishable if one considers *only* these statistics. In principle, the sequences may be empirically distinguished by looking for subtle interference effects between superposed states of correlated systems involved in the chain. For example, the mixture $\Sigma |c_i|^2 \mathbf{P}_i$ (where $\mathbf{P}_i$ projects onto $\phi_i \otimes \chi_i^1$) predicts different statistics for certain correlated variables on O, A_1 than does the superposition $\Sigma c_i(\phi_i \otimes \chi_i^1)$, where A_1 is the first intermediate system in the chain. But in the weak Copenhagen interpretation, which lacks a satisfactory solution to the characterization problem, we do not know which statistics quantum mechanics predicts. And even if we were able to specify the actual link in the chain at which projection occurred, there would remain the problem of saying at what precise *time* during that interaction the time-dependent Schrödinger equation ceased to apply. Von Neumann's "consistency proof" does not remove the imprecision in the weak Copenhagen interpretation: If anything, it indicates that if one accepts the weak Copenhagen interpretation quantum mechanics may be irremediably imprecise!

This conclusion is further reinforced if one considers whether quantum state reduction may be given a relativistically invariant characterization. In the weak Copenhagen interpretation, reduction of the quantum state is understood to be a physical process, accompanied by corresponding changes in the dynamical states of all systems associated with that quantum state. Moreover, there is supposed to be some definite time at which this process occurs. But all systems associated with a given quantum state may not be localized in the same spatial region – consider, for example, an energetic electron-positron pair in a singlet spin state after γ-ray annihilation. Any changes in the dynamical properties of systems that are not localized in the same spatial region can occur simultaneously in only a single inertial frame; and any precise specification of the time at which the joint quantum state is reduced so that these changes simultaneously occur would therefore introduce such a preferred inertial frame. But no relativistically invariant laws describing any physical process may single out a preferred inertial frame: This is the essential content of the special principle of relativity. Hence, there can be no relativistically invariant laws governing the physical process of state reduction postulated by the weak Copenhagen interpretation.

It is important to appreciate exactly why, in contrast, the changes of dynamical state postulated by the present interactive interpretation are in conformity with the requirements of relativistic invariance. For the present interactive interpretation also involves changes in the dynamical states of systems whose components may not be localized in the same spatial region. The critical difference is that, in the present interpretation, the dynamical state of a compound system does not supervene on the dynamical states of its components. Consequently, a change in the dynamical state of a compound system may occur without corresponding simultaneous changes in its components, and so without singling out one spacelike hyperplane on which these changes occur simultaneously. One might say that the time at which a change in the dynamical state of the compound occurs depends on a choice of spacelike hyperplane. But this involves neither inconsistency nor failure of relativistic invariance, since the change consists of variations of irreducible properties across *each* of an infinite set of spatial hyperplanes, each one of which is an element of a different family of parallel hyperplanes – the simultaneity "slices" of an associated inertial frame. In this way the holism of the present interactive interpretation ensures that, by contrast with the weak Copenhagen interpretation, dynamical changes associated with measurement (as well as other interactions) may be given a relativistically invariant characterization.

A comparison with the strong version of the Copenhagen interpretation needs to be made separately, since several of these last detailed criticisms of the weak Copenhagen interpretation proceeded from assumptions that form no part of the strong Copenhagen interpretation. Chief among these is the assumption that a system is always associated with a quantum state which determines its dynamical properties. In the strong Copenhagen interpretation, the sole role of the quantum state is to yield statistical predictions for the possible outcomes of measurements on elements of an ensemble in well-defined experimental conditions: There is no connection between the quantum state of a system and its dynamical state. Consequently, one cannot validly infer that a system described by a superposed quantum state has no definite value for the corresponding dynamical variable. And so, the measurement problem no longer presents the same threat to the empirical adequacy of quantum mechanics, that is averted only by postulating the occurrence of a physical process of quantum state reduction upon measurement. Moreover, for the strong Copenhagen interpretation, insofar as quantum state reduction is not a physical process, there is no need to give a precise frame-independent characterization of when it occurs. Finally, the preparation problem looks very different under the strong Copenhagen interpretation. Since a quantum system is associated with a quantum state only in a well-defined experimental arrangement, the universe is *never* associated with a quantum state, since there is no external apparatus to define such an arrangement. The assumption that one can associate a quantum state with a restricted subsystem of the universe is therefore not open to justification as an approximation. Instead, it is simply taken as a basic assumption of quantum mechanics, in the strong Copenhagen interpretation, that in certain well-defined experimental conditions a quantum state is associated with a restricted subsystem of the universe (or, more precisely, with elements of an ensemble of similar systems).

But this now makes clear the nature of the difficulty faced by the strong Copenhagen interpretation. The problem is to state exactly which sets of conditions constitute "well-defined experimental arrangements," and to show how it is possible to give a consistent, relativistically invariant prescription for associating quantum states with systems in such conditions. It is necessary to solve this problem if the strong Copenhagen interpretation is to render quantum mechanics a precisely formulated theory which permits a precise and nonproblematic account of the role of measurement in quantum mechanics. Note that there are two aspects to the problem. First it is necessary to specify just what constitutes an experimental arrangement. But then it is also necessary to show that in a given experimental arrangement one can consistently associate a quantum state with some system(s), while treating other elements of the arrangement as apparatus, and yield statistics whose recorded results are in accord with those predicted by the Born rules as applied to that quantum state.

Presumably an experimental arrangement exists only if there is at least the possibility of using it to perform a quantum measurement. Consequently, one cannot hope to handle the first aspect of this problem without saying at least approximately when a measurement occurs: This is the analog, in the strong Copenhagen interpretation, of the characterization problem for the weak Copenhagen interpretation. But even if one knew just what constituted a well-defined experimental arrangement, there would remain the difficulty of analyzing it into object system (to be assigned a quantum state) and apparatus system (not to be described quantum mechanically).

To illustrate this second difficulty, consider the following schematic example. Let X be a well-defined experimental arrangement including object system O and apparatus system A. Add to X a further apparatus system B, designed to observe the recording state of A. Presumably, O, A, and B are included in a larger experimental arrangement X'. With respect to X', one can represent the interaction between O and A quantum mechanically. If the interaction between O and A takes the form of an ideal measurement, then, for an initially superposed state of O, the final quantum state of $O \oplus A$ will be an entangled superposition:

$$\Sigma c_i \phi_i \rightarrow \Sigma c_i (\phi_i \otimes \chi_i^A).$$ (6.2)

One can then regard this superposition as predicting the statistics for possible final states of B, following a measurement interaction of B on A.

Alternatively, one might consider $A \oplus B$ to be a complex apparatus, whose possible final states record the result of a measurement interaction on just O. In that case, one will assign a quantum state to O, and treat A as part of the apparatus, not to be described quantum mechanically. Both alternative analyses seem equally legitimate. Moreover, *provided* that the O,A interaction satisfies (6.2), both alternatives predict the same statistics for observed final states of B. It may therefore seem unnecessary to single out one alternative as offering the *correct* analysis of X' into quantum system and apparatus. But note that this is so only on the condition that (6.2) holds. If (6.2) were to fail, it would simply be incorrect to employ the second alternative: A would have to be described quantum mechanically. And this shows that the first analysis, in fact, has priority: It is necessary to employ the first analysis in order to assess the adequacy of the second analysis. Pursuing this line of reasoning would lead one to conclude that the more of an experimental arrangement like X' that is treated quantum mechanically, the more exact is the quantum mechanical analysis of X'. But there is no natural limit to a process of treating more and more of X' quantum mechanically, short of describing the whole of X' by a quantum state – in which case X' no longer contains any apparatus system, the statistics for whose final states this quantum state may predict! In the strong Copenhagen interpretation, there is no nonarbitrary way of analyzing a given experimental arrangement into object system and apparatus, even though the quantum mechanical predictions concerning that arrangement may well depend on exactly how that division is made. Hence, the strong Copenhagen interpretation also seems irremediably imprecise.

Moreover, the strong Copenhagen interpretation also fails

to account for the great explanatory power of quantum mechanics. If an application of quantum mechanics were to require a specification of the exact experimental arrangement, then quantum mechanics could not be applied outside of these arrangements. But quantum mechanics is widely believed to be capable of explaining the behavior of matter at both microscopic and macroscopic levels; and though such explanations do involve detailed assumptions about the structure and state of this matter, there is typically no correspondingly detailed description of the experimental arrangements in which quantum mechanics is being applied. This suggests that, contrary to the strong Copenhagen interpretation, quantum mechanics may be used to describe and explain phenomena involving quantum systems independently of the details of any experimental arrangements that may be employed to study these phenomena.

Even with a detailed specification of the experimental arrangement, quantum mechanics lacks explanatory power under the strong Copenhagen interpretation, both because of its silence on the evolution of dynamical properties of systems, and because of its lack of precision concerning the ascription and evolution of quantum states. This is nicely illustrated by the example of the Bohm version of the EPR thought-experiment described in Chapter 4. For how, in the strong Copenhagen interpretation, can quantum mechanics account for the observable correlations in this experiment? Although quantum mechanics does indeed correctly predict their occurrence, adoption of the strong Copenhagen interpretation effectively precludes the theory from explaining how they come about. Any such explanation would proceed by ascribing states to the pairs of systems concerned, and tracing their evolution through the measurements which display each observable correlation. But, in the strong Copenhagen interpretation, quantum mechanics does not itself ascribe dynamical states to quantum systems: Instead, it *assumes* that a quantum measurement will produce some appropriate dynamical state of one or more apparatus systems, and then yields probabilistic predictions concerning what such states will be. It is true that for the strong Copenhagen interpretation quantum mechanics ascribes a quantum state to elements of an ensemble of pairs in the Bohm version of the EPR thought-experiment, prior to any spin measurements on the pairs. But as already noted, this ascription has no implications concerning the initial dynamical properties of the pairs. Moreover, as it stands, the strong Copenhagen interpretation offers no precise account of how even the *quantum* states of the pairs evolve during the course of the two spin measurements to which each pair is subjected.

Now it is possible to supplement the strong Copenhagen interpretation by describing how the quantum states of these pairs evolve. But this in no way helps to explain how the observed correlations come about. That description is the following. After only the first spin-component measurement, say of z-spin on system A, the quantum state of B is determined as a factor in the result of applying the projection postulate to the initial quantum state ψ^i_{AB} of the pair. Thus, for an initial singlet state, we have

$$\psi^i_{AB} = (1/\sqrt{2}) \left[(\zeta^+_A \otimes \zeta^-_B) - (\zeta^-_A \otimes \zeta^+_B) \right] \qquad (6.3)$$

$$\xrightarrow{\text{projection}} \zeta^+_A \otimes \zeta^-_B,$$

from which we conclude that the quantum state of B elements of the pairs that give spin-up for the measurement of z-spin on A is the factor ζ^-_B of $\zeta^+_A \otimes \zeta^-_B$, where $\zeta^\pm_{A\ (B)}$ is an eigenstate of z-spin of the A (B) system with eigenvalue $\pm \hbar/2$. This accords with the conditional certainty of obtaining a result $-\hbar/2$ for a measurement of z-spin on a B system whose corresponding A system yielded a result $+\hbar/2$ for the corresponding measurement. Indeed, ascription of the state ζ^-_B to such B elements yields, via the Born rules, the same probabilities for *all* measurement results on these elements as are obtained by applying the Born rules directly to the preprojection quantum state of (6.3) and conditionalizing on the $+\hbar/2$ result of the A measurement. Hence, we have a consistent account as to how the first A measurement is accompanied by alterations of the quantum states of the systems concerned. Prior to the first A measurement, all pairs of the original ensemble have the (same) singlet quantum state, whereas their component systems have no distinct quantum states: After the first A measurement, the B systems divide into two subensembles, with quantum states ζ_B^- and ζ_B^+, according to whether the result of the z-spin measurement on the corresponding A system was $+\hbar/2$ or $-\hbar/2$. Moreover, this account is readily extended to the symmetric case in which it is the B measurement which occurs first.

There is indeed a problem rendering the account relativistically invariant: In particular, if the A and B measurements on each pair are spacelike separated events, they have no invariant time ordering. But even this problem may be overcome in the manner of Fleming (1985) by making all quantum state ascriptions hyperplane-dependent. Thus, if the M_A measurement is confined to space-time region R_A, and the M_B measurement is confined to (nonoverlapping) region R_B, then a generalization of (6.3) ascribes a quantum state to a system B on any spacelike hyperplane that intersects only the causal future of R_A and only the causal past of R_B, and a quantum state to A on any spacelike hyperplane that intersects only the causal future of R_B and only the causal past of R_A. But though this may render the account relativistically invariant, it also makes it particularly clear that such quantum state ascriptions to A,B systems play no role in explaining how the observable correlations come about. Ascribing a quantum state to a system cannot be ascribing any real property to that system if whether or not the system has that quantum state may depend not only on the space-time region in which the system is located, but also on how an arbitrarily chosen hyperplane is located with respect to that region.

To conclude this comparison with the Copenhagen interpretation, let me contrast its account of a classic example of quantum measurement with that offered by the present interactive interpretation. The example is provided by the use of a Stern–Gerlach device to measure spin. Its theoretical analysis in the present interpretation gives a concrete application of Chapter 3's general account of measurement, and also makes it clear how the present interpretation escapes the negative conclusion of what may be called the **backtracking argument**,[5] a line of reasoning closely related to the measurement problem.

When a Stern–Gerlach device is used to measure the spin-component of a monochromatic beam of neutral particles of some kind, each particle passes through an inhomogeneous magnetic field which varies along the chosen spin axis, and the particles are subsequently detected at various locations along this axis. Suppose, for simplicity, that the particles in the beam have spin-½. Then every particle in an incoming beam described by a quantum state which is an eigenstate of spin-component in the z-direction with eigenvalue $+\hbar/2$ ($-\hbar/2$) will (ideally) subsequently be detected in a region of positive (negative) z-coordinate. In the weak Copenhagen interpretation, each particle in either of these incoming beams has a definite value of z-spin: $+\hbar/2$ for each particle in one beam and $-\hbar/2$ for each particle in the other. These values are simply revealed by detection in the relevant region of the z-axis after passage through the magnetic field. An incoming beam described by a nontrivial superposition of these spin eigenstates will be split, with some random selection of particles detected in the region of positive z-coordinates, and a different random selection detected in the region of negative z-coordinates. In the weak Copenhagen interpretation, though none of the particles in the original beam had a definite

---

[5]I owe this name to Geoffrey Hellman, who brought this argument to my attention as presenting a potential problem for the present interpretation.

value of z-spin, passage through the magnet appears to have produced two distinct outgoing beams, with each particle in the upper (lower) beam having z-spin $+\hbar/2$ ($-\hbar/2$). It then seems that passage through the magnet itself must constitute a measurement which randomly projects the superposed state of each particle onto one or other of its component z-spin eigenstates. This conclusion seems further reinforced by noting that every particle in the hypothetical upper beam emerging from the Stern-Gerlach magnet will be deflected upward once more on passage through a second similar Stern-Gerlach magnet also oriented in the z-direction, whereas all particles in the hypothetical lower beam will be deflected downward by such a second Stern-Gerlach magnet. Thus, it appears that, in the weak Copenhagen interpretation, passage through a Stern-Gerlach magnet itself constitutes a measurement of spin-component, accompanied by projection of the quantum state of the particles concerned.

But this conclusion has been objected to, on different grounds, by both Margenau and Wigner.[6] Margenau objects that no measurement has occurred until a particle is actually detected in some region of the z-axis after passing through the Stern-Gerlach magnet, for only then does the device record any measurement *result*. According to Margenau, passage through the Stern-Gerlach magnet merely prepares the quantum states of outgoing particles, with particles in the upper (lower) beam described by the positive (negative) eigenvalue eigenstate. In his view, whereas such a state preparation may be described by the projection postulate, a typical measurement may not be so described: In this case, the only direct measurement is of the particles' positions, and there is no reason to suppose that this produces a position eigenstate. Wigner, on the other hand, denies that any measurement occurs on passage through the Stern-Gerlach magnet, on the grounds that whereas a genuine measurement is characterized

[6]See, for example, Margenau (1963) and Wigner (1963).

by projection, no projection has yet occurred. He supports this contention by appeal to a thought-experiment in which the two outgoing "beams" are recombined and made to interfere, with observable results that are incompatible with those produced by the quantum states resulting from application of the projection postulate. Though Wigner's recombining beams remain a thought-experiment, the results of analogous experiments with split-beam neutron interferometers lend experimental support to his contention that passage through the Stern-Gerlach magnet does not bring about a distinctive physical process of projection.[7] But if Wigner's contention is correct, then it becomes difficult to understand how a Stern-Gerlach device can be used to measure spin, or even to prepare eigenstates of spin as Margenau claims.

The objections of Margenau and Wigner leave one with the following puzzles. Does passage through a Stern-Gerlach magnet constitute a measurement of spin-component? And, if not, how can a Stern-Gerlach device be used to measure spin-component? Can a Stern-Gerlach device be used to prepare quantum eigenstates of spin-component? And, if so, how is this possible? These questions receive the following answers in the present interactive interpretation.

Consider a double Stern-Gerlach arrangement in which a monochromatic beam of neutral particles of the same kind is first incident on a Stern-Gerlach magnet whose axis is oriented along the z-direction, and then a second Stern-Gerlach magnet with axis oriented along a different z'-axis is placed to intercept just one of the (hypothetical) beams which emerge from the first Stern-Gerlach magnet. In this repeated Stern-Gerlach experiment, interaction with the first magnet is part of a preparation procedure which renders legitimate the ascription of a $+\hbar/2$ eigenstate of z-spin to those particles whose z'-spin is subsequently measured by detecting their passage through the second Stern-Gerlach magnet. This il-

[7]A nice review of some such experiments is given by Greenberger (1983).

lustrates the general point that, by itself, passage through a Stern–Gerlach magnet constitutes neither a measurement nor a state preparation: But passage through a Stern–Gerlach magnet may constitute a critical element in either a spin-state preparation procedure or a measurement of spin-component, depending on the physical context within which such passage occurs.

In the double Stern–Gerlach arrangement just described, passage through the first Stern–Gerlach magnet is an element of a type-2 simple preparation procedure, which also features interaction with the environment, including (in particular) interaction with the detector following passage through the second Stern–Gerlach magnet. The final element in this preparation procedure consists of the selection of that subset Σ' of particles from the original beam which is detected following passage through the second Stern–Gerlach magnet: It is this subset of particles which constitutes the ensemble to which the $+\hbar/2$-spin eigenstate may legitimately be ascribed in the interval between the two Stern–Gerlach magnets.

In this same arrangement, passage through the second Stern–Gerlach magnet plays a role in a measurement of the z'-spin of particles in the ensemble Σ'. That role is somewhat indirect, since the second Stern–Gerlach magnet does not act as the apparatus system in a measurement-type interaction with the particles passing through it: The apparatus system is rather the detector which records the presence of particles in Σ' in one of two regions of the z'-axis after passage through the second Stern–Gerlach magnet. The second Stern–Gerlach magnet serves rather to effect a correlation between the (internal) spin state of particles incident upon it and their (external) spatial state. This is necessary insofar as the detector interacts with the particles only through their spatial state and is insensitive to their spin. It is interesting to note that an interaction with a Stern–Gerlach magnet would have the correct mathematical form to be suitable for an M-type interaction, if one were to suppose that each particle is composite, with one subsystem describing its spin state and the other component subsystem describing its spatial state. But this supposition is in fact false, and this is the basic reason why, in the present interpretation, the interaction with a Stern-Gerlach magnet does not itself constitute a measurement.

The present interpretation therefore endorses Margenau's position that the only measurement interaction occurs when a particle is detected after passage through the magnet, and that this need not produce a position eigenstate. But it differs from Margenau's view in maintaining that passage through a Stern-Gerlach magnet is only the first of two interactions required to prepare a quantum spin state. The preparation is completed only by a further interaction with the detector (or other part of the environment) following passage through the magnet. This makes the preparation in a sense retrospective: It is only because of an interaction that a particle will experience when it is detected that it can legitimately be ascribed a quantum spin state during the interval when it is between the two Stern-Gerlach magnets. But this should not be surprising if one recalls the phenomenon of the delayed-choice experiment, and how such experiments are understood in the present interpretation.[8] As for Wigner's view, the present interpretation agrees that no measurement has occurred just after passage through (either) Stern-Gerlach magnet, and that recombining beams would lead to interference effects. But it explains how passage through a Stern-Gerlach magnet can nevertheless form an important element in a measurement of spin-component, provided that it is followed by a measurement-type detection interaction (specified in purely physical terms!). And it also explains how passage through a Stern-Gerlach device can form an important element in a preparation procedure, even though it does not induce projection onto a single component spin eigenstate. Here, it is important to remember that there is no inconsistency between assigning

[8] See the discussion in Chapter 1, p. 37.

a spin eigenstate to an ensemble of particles emerging from a Stern-Gerlach magnet, and maintaining that if instead of being detected these same particles had been subjected to interactions designed to "recombine the beams" prior to detection, they would have revealed interference effects which would not have been predicted by that spin eigenstate. Perhaps the best way to see that there is no inconsistency is to note that any interaction with the environment necessary to complete state preparation following passage through the Stern-Gerlach magnet precludes the interactions necessary to recombine the beams. Therefore, if a state is in fact prepared, no interference effects will be demonstrable, and vice versa.

The back-tracking argument is the following. Suppose (as does the present interpretation) that a measurement interaction leaves both the apparatus system and the object system in a definite state. If a system is in a definite state, then it must be in an eigenstate of some dynamical variable. Hence, if the same measurement is performed on an ensemble of similar object systems by a collection of similar apparatus systems, each of the apparatus systems and corresponding object systems must be in an eigenstate following the measurement interaction. It follows that each object system/apparatus system pair must also be in an eigenstate, namely, the direct product of their individual eigenstates. Consequently, the postmeasurement ensemble must be described by a mixture of such direct product states, each with appropriate statistical weight; and in the typical case these weights will be such that this mixture is not a pure case. But if the measurement interaction proceeds in accordance with the time-dependent Schrödinger equation, there must be some unitary transformation which gives the final state when applied to the initial state: And any unitary transformation applied to a pure case gives another pure case. Hence, the premeasurement ensemble cannot have been described by a pure case. But this contradicts the assumed initial condition, that the ensemble was indeed initially described by a pure case, namely, a direct product of a pure apparatus state and a pure superposed system state. By reductio ad absurdum it follows that a measurement interaction cannot leave both system and correlated apparatus in definite states. Therefore, the present interpretation must be rejected.

One may attempt to make this abstract argument more concrete by considering a measurement of spin-component in the z-direction performed by means of a Stern-Gerlach device on a polarized beam of particles with spin in some nonparallel z'-direction. The argument may then be used to conclude that the measurement can give a definite result for each particle only if the Stern-Gerlach magnet projects each particle's spin state onto an eigenstate of z-spin, in violation of the time-dependent Schrödinger equation. This concrete form of the argument is refuted by noting that even though in the present interpretation passage of each particle through the Stern-Gerlach device yields a definite result, it does not follow (and nor is it true) that each particle is projected onto an eigenstate of z-spin by passage through the Stern-Gerlach magnet. It is true that passage of a particle through the magnet of a single Stern-Gerlach device does, in this situation, help to prepare an eigenstate of z-spin. And it is also true that the detector is in a definite state when it records the position of the particle which has passed through the magnet. But it does not follow that the detector may be described by an eigenstate of its recording quantity after detecting the particle, since no preparation procedure for such an eigenstate has been specified. And even if the detector were then legitimately described by an eigenstate of its recording quantity, it would not follow that there was a time when a particle/detector pair was correctly described by a product state of which this was one component. In fact, there is no such time. After the particle/detector interaction, these systems have become correlated in such a way that even if a pair could then legitimately be described by a quantum state, it would be a state which would imply predictions concerning joint measurements on detector and particle which would conflict with the predictions of any product quantum state.

Returning now to the abstract form of the argument, it is clear that there are two distinct steps in the argument which are mistaken. In the present interpretation, it does not follow from the fact that a system is in a definite dynamical state (i.e., has a definite value for each of some complete, commuting set of dynamical variables) that that system may be described by a pure quantum state: The connection between quantum states and dynamical states is less direct than that. Nor does it follow from the fact that each of two systems may be described by a quantum state that the joint system that they compose may be described by the product of these quantum states. An inference to the product state is legitimate only in certain circumstances. Crudely, the product rule holds only when the states of the components are uncorrelated: Some more precise restrictions may be determined by considering the specifications of preparation procedures and measurement interactions offered in Chapter 3.

§6.3 EVERETT

Of all the better-known interpretations of quantum mechanics, it is Everett's which has proven most influential in the development of the present interactive interpretation. This is why it is particularly appropriate for me to compare the two here, and to make clear why I regard the present interpretation as a significant improvement on Everett's.

The Everett interpretation may be regarded as the prototype of all interactive interpretations, since it was perhaps the earliest and most influential attempt to treat measurement as a physical interaction internal to a compound quantum system, one component of which represents the observer or measuring apparatus. The Everett interpretation, like the present interactive interpretation, rejects the projection postulate. Both interpretations maintain that all interactions, including measurement interactions, may be treated as internal to a compound system, the universe, whose state evolves always in accordance with a deterministic law such as the time–dependent Schrödinger equation. And both deny that it is necessary to appeal to any extraquantum-mechanical notions such as that of a classical system, or an observer, in order to give a precise and empirically adequate quantum mechanical model of a measurement interaction. Finally, both interpretations undertake to explain how, and to what extent, quantum interactions internal to a compound system can come to mimic the effects of the projection postulate, even though no such postulate forms part of quantum mechanics.

A more fine-grained comparison with the Everett interpretation will require a closer analysis of that interpretation itself – one that is capable of importantly distinguishing different ways of developing Everett's original ideas.[9] But it is useful at the outset to recall Everett's own strategy for reconciling his denial that there is any such physical process as projection with his acceptance that measurement results appear to accord with the occurrence of such a process. According to Everett,[10] all observers correspond to quantum systems, which may be called, for convenience, **apparatus systems**: An observation or measurement is simply a quantum interaction of a certain type between an apparatus system α and an object system σ, which (provided this compound system is isolated) proceeds in accordance with the time-dependent Schrödinger equation governed by the Hamiltonian for the pair of systems concerned. In particular, for a **good observation** of a dynamical variable $\mathcal{A}$ whose associated operator $\mathbf{A}$ has complete set of eigenvectors $\{\phi_i\}$, the interaction Hamiltonian is such that the joint quantum state immediately after the conclusion of the interaction is related to the initial state as follows:

[9]See Healey (1984a, 1984b).
[10]See his "The Theory of the Universal Wave Function," and "'Relative State' Formulation of Quantum Mechanics," reprinted in DeWitt and Graham (1973).

$$\psi^{\sigma \oplus \alpha} = \phi_i^\sigma \otimes \psi_{[\ldots]}^\alpha \rightarrow \psi'^{\sigma \oplus \alpha} = \phi_i^\sigma \otimes \psi_{i[\ldots,a_i]}^\alpha \qquad (6.4)$$

for each eigenvector ϕ_i^σ of **A**, where the ψ_i^α are orthonormal vectors, $[a_i]$ stands for a recording of the eigenvalue a_i of **A**, and the dots indicate that results of earlier good observations may also be recorded in the state of α. It then follows from the linearity of the time-dependent Schrödinger equation that an arbitrary (normalized) initial object system quantum state $\Sigma c_i \phi_i^\sigma$ (with $\Sigma |c_i|^2 = 1$) gives rise to the following transformation:

$$\sum_i c_i \phi_i^\sigma \otimes \psi_{[\ldots]}^\alpha \rightarrow \sum_i c_i (\phi_i^\sigma \otimes \psi_{i[\ldots,a_i]}^\alpha). \qquad (6.5)$$

For Everett himself, each component $\phi_i^\sigma \otimes \psi_{i[\ldots,a_i]}^\alpha$ with non-zero coefficient c_i in the superposition on the right-hand side of (6.5) corresponds to a distinct state in which the observer has recorded the ith eigenvalue for the measured quantity on the object system, while the object system remains in the corresponding eigenstate ϕ_i. Moreover, all these states are equally real – every possible result is recorded in some observer state $\psi_{i[\ldots,a_i]}^\alpha$, and there is no unique actual result.

For a sequence of good observations by a single observer, consisting of multiple pairwise interactions between the apparatus system and each member of a set of object systems, Everett is then able to show the following. If a good observation is repeated on a single object system in circumstances in which that system remains undisturbed in the intervening interval (in the sense that the total Hamiltonian commutes with the operator representing the observed quantity), then the eigenvalues recorded for the two observations are the same, in every observer state: I shall call this the **subjective reproducibility of measurements (SRM)**. Note that this is exactly what would be predicted by an observer who represented each object system independently by a quantum state vector and regarded the first of each sequence of repeated measurements on it as projecting the relevant object system's

207

quantum state onto an eigenvector corresponding to the initially recorded eigenvalue: This is the first respect in which, for each observer, a good observation *appears* to obey the projection postulate.

Everett next shows that each observer will get the right *probabilities* for results of arbitrary good observations on a system which has been subjected to an initial good observation, if, following this initial observation, he assigns to the system the quantum state it would have had if projection had then occurred. For the following two probabilities are demonstrably equal: the probability of result b_j in a subsequent good observation of $\mathcal{B}$ on σ by an observer corresponding to α who applies the projection postulate to the state of σ alone after an initial good observation of $\mathcal{A}$ on σ by α yielding result a_i; and the probability (assuming that the state of the compound $\sigma \oplus \alpha$ evolves according to the time-dependent Schrödinger equation) that after the $\mathcal{B}$ measurement the observer state of α will record the values a_i and b_j, conditional on the observer state of α after the $\mathcal{A}$ measurement recording the result a_i of the initial observation. This demonstration explains how, for each observer, it is as if a good observation prepared a corresponding eigenstate of the observed system. But this still does not suffice to establish that everything is as if projection actually occurs. There are two further consequences of projection that need to be mimicked.

If projection really occurred, then each of several independent observers performing repeated good observations of the same quantity on an otherwise undisturbed system would necessarily obtain the same result. I shall call this the requirement of the **intersubjective reproducibility of measurements (IRM)**. Moreover, if projection really occurred, then the state of α immediately after a good observation would be one of the ψ_i^α appearing on the right-hand side of (6.5). In this apparatus state the "pointer position" quantity has its ith eigenvalue, recording that the observed quantity had its *i*th eigenvalue on σ. And in this apparatus state the probability is 1 that a subsequent observation of the pointer position quantity would reveal that it has its *i*th eigenvalue. Consequently, the result of a subsequent observation of the "pointer position" quantity on an undisturbed apparatus system will reveal the value that the "pointer position" quantity had at the conclusion of the initial interaction with σ – the value which recorded the result of the measurement on σ. I shall call this consequence of projection the condition of the **verifiability of measurement results** (**VMR**).

Now, Everett does not (and cannot) show that IRM and VMR hold for his interpretation. Instead, he shows that IRM and VMR *appear* to hold for any (good) observer, in the sense that each observer state will come to contain records consistent with both holding, and there is no chance of its coming to contain records of their failing to hold. Thus, VMR appears to hold since (for example) if an observer corresponding to α_1 repeatedly makes good observations of the pointer position quantity on an undisturbed apparatus system α_2 which has itself made a good observation on σ, then (provided that these are the only interactions involving the systems) the probability that the observer state of α_1 comes to record the *j*th eigenvalue for the second observation, following a recording of the *i*th eigenvalue for the first observation, is δ_{ij}. Two examples serve to illustrate why IRM also appears to hold, in this sense. Suppose that each of α_1 and α_2 performs good observations of the same quantity $\mathcal{A}$ on σ, and then α_1 performs a good observation of the pointer reading quantity $\mathcal{B}$ on α_2. Provided that the three systems are otherwise undisturbed, it is certain (with probability 1) that the observer state of α_1 will come to record values for $\mathcal{A}$ and $\mathcal{B}$ that correctly correspond to one another. Suppose, on the other hand, that each of α_2, α_3 performs good observations of $\mathcal{A}$ on σ, and then α_1 performs good observations of the pointer reading quantities $\mathcal{B}_2, \mathcal{B}_3$ on α_2, α_3, respectively. Then, with the analogous proviso, the observer state of α_1 will certainly (probability 1) come to record values for $\mathscr{B}_2, \mathscr{B}_3$ which each correspond to the same value of $\mathscr{A}$.

One might object that an observer does not need to perform an observation on another observer in order to find out the result of the latter's observation – one only has to ask (or more generally communicate in some way). Everett's reply is that all observers correspond to apparatus systems, that all communication with an observer involves some interaction with the apparatus system that corresponds to that observer, and that good observations are the *least* disturbing interactions permitting the necessary communication, and consequently serve as an idealized model for all actual communications.

Despite the elegance of Everett's strategy for mimicking the effects of the projection postulate while maintaining that the quantum states of all isolated systems evolve deterministically and unitarily, his interpretation is subject to a fundamental objection.[11] The objection is that it fails to satisfactorily explain how an observer can observe a definite outcome of any individual quantum measurement on a system in a nontrivially superposed state. Everett says that each element of the superposition on the right-hand side of (6.5) represents an observer state which records a different result of the observation. But if one performs a quantum measurement, one always observes that the apparatus ends up in *one* definite state (e.g., the pointer points at some definite position on the dial): Prior to the measurement other outcomes were indeed possible, but one actually occurred whereas the rest did not. A proponent of the Everett interpretation must try to show that his view is neither inconsistent nor empirically inadequate. The view seems inconsistent insofar as it claims that following a good observation an observer may be in incompatible states; and it seems empirically inadequate in-

[11]See Healey (1984a, 1984b); Stein (1984).

sofar as it fails to accommodate the fact that just one of these states is actualized.

Some supporters of Everett have tried to combat the charge of inconsistency by supposing that physical systems literally fission in a measurement interaction. According to this view, each quantum measurement interaction splits each quantum system (and perhaps even physical space as well) into multiple copies of itself. The copies sort themselves into worlds, so that (for example) in one world emerging from an interaction represented by (6.5), σ is in state ϕ_1 and α is in state ψ_1 (recording the definite result a_1), while in another world σ is in state ϕ_2 and α is in state ψ_2 (recording the definite result a_2). Moreover, these worlds are not merely possible, but equally real. I have argued elsewhere that the fantastic ontological cost of this view is not balanced by any corresponding explanatory benefit, as compared to a view according to which all but one of the worlds emerging from a quantum measurement are merely possible, whereas there is a single actual world in which the observer records a definite result (it is then still open to the modal realist to try to argue that the excess possible worlds are themselves real, though not actual).[12] Indeed, as I shall explain, the present interactive interpretation originated as an attempt to work out the details of a defensible version of such a "one-world" construction of the Everett interpretation. But there is one other version of the Everett interpretation that needs to be considered first, namely, Everett's own!

As I have explained elsewhere, I believe that Everett himself is best interpreted as holding that the correspondence between observers and apparatus systems may be many-to-one rather than one-to-one, and that an observer is to be identified not with any physical system, but rather with a physical *process* associated with an apparatus system, and constituted by a

[12]See Healey (1984a).

sequence of relative (observer) states of that system.[13] It is then not physical systems but physical processes which branch in quantum measurement: Each emerging process incorporates a different relative observer state, and each different relative observer state records a different result of the measurement. There is no absolute observer state after the measurement interaction. One advantage of this view over the many-worlds view is that the branching it postulates is restricted only to those entities involved in the measurement itself, and does not require that every physical system (or physical space itself) should split in a quantum measurement. But this view is still ontologically extravagant, postulating the (frequently infinite) reduplication of entities in quantum measurements. And indeed these entities are even of a new ontological category: They are processes which do not consist of sequences of (absolute) states of objects. Even one of Occam's duller razors will eliminate these entities, provided that a defensible one-world version of the Everett interpretation can be developed.

But can this be done? One perspective on the present interactive interpretation is that it is an attempt to develop just such a defensible version. Though ontologically conservative, such a view nevertheless faces a number of important problems. Perhaps most important are the problem of specifying the nature of definite states, and the problem of saying precisely when, and under what conditions, an interaction produces a definite state in a system which undergoes it.

Everett identifies a definite state with a relative quantum state of a system, where a relative quantum state is a factor of a component from a superposition which expresses the quantum state of the universe. Such a definite state specifies those dynamical properties which have probability 1 when the Born rules are applied to the given (relative) quantum state. It is not clear whether Everett himself supposes that

---

[13]See Healey (1984b).

every interaction produces a splitting in processes, that only measurement interactions do so (he gives no precise quantum mechanical definition of a general measurement interaction), or that it is only good observations which produce splitting. The most natural development of his ideas may be to assume that *every* interaction can result in an indeterministic transition to a definite state of *any* quantum system involved in that interaction: This is just what the present interactive interpretation assumes. But this assumption gives rise to the following difficulty.

One can show that an arbitrary normalized state vector in the product space corresponding to a compound system $\sigma \oplus \tau$ may be expanded in the form

$$\psi = \sum_i c_i (\phi_i^\sigma \otimes \phi_i^\tau), \qquad (6.6)$$

where the ϕ_i^τ are selected from a *chosen* orthonormal basis, but the ϕ_i^σ need not be orthogonal. Hence, vectors from an *arbitrary* basis may be selected to appear as relative states for τ in the expansion of the quantum state of a compound system $\sigma \oplus \tau$ following an interaction with σ. In the one-world version, just one vector ϕ_i^τ from such a basis represents the actual definite state of τ following the interaction (with probability $|c_i|^2$. But it is also necessary to specify some nonarbitrary way of choosing the basis from which this vector is selected. Moreover, this must be done in a way such that if τ is an apparatus system α, then the basis is such that the definite state of α following the interaction specifies dynamical properties corresponding to a definite value for the pointer position quantity on α. This requires some satisfactory specification of the class of measurement interactions, and a demonstration that such interactions do indeed give rise to definite values for pointer position quantities. Finally, the condition for an interaction to have concluded must be precisely specified. Note that if all the required specifications are to be satisfactory, they must be given in quantum mechanical terms, without any appeal to problematic undefined external terms such as 'measurement' or 'classical system.'

The present interactive interpretation does indeed offer such specifications in quantum mechanical terms. The subspace decomposition condition effectively removes the difficulty of selecting a preferred basis in (6.6) by appeal to the biorthogonal decomposition lemma: That lemma shows how the expansion in (6.6) becomes essentially unique, if it is required to be biorthogonal, and, hence, picks out a preferred basis solely by appeal to the form of ψ itself. The treatment of measurement given in Chapter 3 shows how it is possible, in the present interpretation, to give a purely quantum mechanical specification of measurement interactions in such a way as to ensure that these do indeed give rise to definite pointer readings (even for nontrivially superposed initial states). And that chapter also offers and defends a quantum mechanical specification of the conditions for such an interaction to conclude. In all these respects, the present interactive interpretation is to be preferred to any one-world version of the Everett interpretation which fails to offer satisfactory solutions to these problems. It seems to be an unimportant residual semantic question as to whether the devices employed by the present interpretation in solving these problems distance the interpretation from Everett's own writings so far that it would be incorrect to consider the present interpretation to be yet another version of the Everett interpretation.

Although it owes much to Everett's strategy for mimicking the effects of the projection postulate, there are significant differences of detail between the present interactive interpretation and Everett's own. Everett shows that SRM holds for good observations, and takes this to be part of what is required to mimic projection. In the present interpretation, the results of measurements are subjectively reproducible only in certain circumstances: These are circumstances in which an initial measurement interaction proceeds in accordance with an interaction Hamiltonian which commutes with the operator representing the measured quantity. This will be so, for example, if the interaction represented by (3.1) is such that **A** commutes with the Hamiltonian during and after the measurement it permits. The stability condition is the key to the subjective reproducibility of measurements in the present interpretation. It is an advantage of the present interpretation that it also allows for measurements whose results are *not* reproducible, since this is true of many actual measurements.

Everett's demonstration that projection yields the right statistics for results of repeated good observations on a system is a model for the present interpretation's derivation of the Born rules by conditionalization. It therefore also holds the key to the present interpretation's treatment of state preparations. But within Everett's own interpretation it proves too much or too little: for not all observations are suitable for preparing states, and not all state preparation interactions yield recordings of an observation.

As for IRM and VMR, whereas Everett himself is able to show only that it appears to all good observers that each of these conditions holds, in the present interpretation one can show that, in certain circumstances, each of these conditions actually holds. Such a demonstration is given for VMR in Chapter 3. Once more, it is the stability condition which is the key to these demonstrations. And once again, the conditions generally hold only in circumstances in which the interaction Hamiltonian commutes with the operator representing the appropriate quantity – either the initially measured quantity (for IRM) or the pointer position quantity (for VMR). It is an advantage of the present interpretation that it allows for nonreproducible and/or nonverifiable measurement results, while permitting an investigation of the exact circumstances in which measurement results will be either reproducible or verifiable.

Notice that whereas the present interpretation postulates the stability condition explicitly, Everett implicitly relies on

215

a similar condition. He implicitly assumes that the "records" in an observer state continue faithfully to represent what actually happened for that observer: that if an observer state records a result of the mth good observation following the nth good observation ($m < n$), then that is both the same result it recorded *before* the nth good observation, and also the result that the mth good observation actually produced for that observer. It is only if these assumptions are granted that it is legitimate to represent the history of a set of "genetically related" observer processes by means of a tree which branches from a single trunk. And it is only if these assumptions are granted that Everett can claim to have successfully mimicked the effects of the projection postulate.[14]

§6.4 KOCHEN

Finally, I shall compare the present interactive interpretation to the views of Kochen, as expressed in an unpublished manuscript, "A New Interpretation of Quantum Mechanics," dating from 1978, as well as in a paper of the same title, but significantly different content (Kochen, 1985). The present interactive interpretation is indebted to the unpublished manuscript, but was almost completed before I became aware of Kochen (1985). In these papers, Kochen presents a novel approach to the measurement problem as well as some interesting ideas on EPR-type correlations. I commented on Kochen's earlier views in 1978, and my comments were published without the paper on which they were based in Healey (1981). At that time it seemed to me that the objections raised against Kochen's views by myself and also by Norman (1981) were sufficiently powerful that it was not fruitful to continue

[14]It is interesting to note that in his critique of the Everett interpretation, Bell (1987, p. 135) objects to just these implicit assumptions. However powerful his objections may be against the views of Everett or DeWitt, I do not see that they have any force against the present interactive interpretation's analogous assumptions (the stability condition, and the correlation constraint built into the subspace decomposition condition).

further work on these ideas. Accordingly, I set them aside until early 1984, and explored the Everett interpretation instead.

Then, while developing a one-world version of the Everett interpretation, I began to reexamine the objections to Kochen's ideas. I came to see that, despite the objections, these ideas suggested a framework within which it might be possible to implement a one-world version of the Everett interpretation. Although I continue to believe that there are powerful objections against Kochen's own original presentation of his views, I then realized that some objections could be satisfactorily answered, and certain significant but natural changes and extensions of the framework suggested by Kochen's ideas could provide an answer to others.[15] It was only when I had satisfied myself that all the objections to Kochen's ideas could be deflected in this way that I felt able to begin serious work on what subsequently became the present interpretation. Meanwhile, Kochen's own thinking took a different turn, and whereas his 1985 paper does contain some revisions that bring his views closer to elements of the present interactive interpretation, there are more important respects in which our views have come to diverge.

I recite this history to clarify the ways in which not only the content but also the ambitions of our interpretations of quantum mechanics differ. Although my thinking is indebted to ideas of Kochen, it is perhaps equally indebted to ideas of Bohr, Bell, and Everett. Perhaps more importantly, whereas in the papers referred to above Kochen has sought to shed light on a few selected problems in understanding quantum mechanics, I have sought to develop an approach which is capable of handling all the outstanding conceptual problems of quantum mechanics. An interpretation will be ultimately acceptable only if it provides a way of thinking about quan-

[15]I owe a special debt here to Al Janis. It was while I was trying to explain to him why Kochen's views were unacceptable that I came to see how they might be transformed so as to meet the criticisms I was making!

tum mechanics which explains why, when understood in the right way, quantum mechanics raises no conceptual problems distinct from the problems science always raises when it alters and extends concepts to account for new phenomena.

The central idea of Kochen's unpublished 1978 manuscript was to suppose that every interaction of finite duration between two otherwise isolated quantum systems results in the two systems' coming to possess correlated sets of dynamical properties, even though the Born rules do not assign probability 1 to these properties given the joint quantum state after the interaction. The technical device to which he appealed to state precisely what these correlated properties are is the polar decomposition theorem – a result essentially equivalent to the biorthogonal decomposition lemma of Chapter 2. Within this framework, the Born rules are understood as specifying the probability that a dynamical property will be realized in a suitable interaction; measurements are considered to be a special case of two-system interactions of finite duration, in which the dynamical properties realized on the apparatus system include the pointer position quantity; and EPR-type correlations are explained (in part) in terms of the properties realized by the two-body interactions that set them up.

The polar decomposition theorem may be viewed as providing a natural selection of a preferred basis for an expansion like (6.6), thus contributing to the solution of an outstanding problem for the one-world version of the Everett interpretation. In this role, it is incorporated also into the present interactive interpretation, through the subspace decomposition condition. However, there are significant differences between the structure of dynamical states postulated by the present interactive interpretation and the dynamical properties Kochen takes to be realized following an interaction. In particular, Kochen recognizes no distinction between reducible and irreducible properties, whereas this distinction is crucial both to the treatment of measurement interactions and to the explanation of EPR-type correlations given by the present interactive interpretation. In both of these respects the present interpretation is superior to Kochen's 1978 views, or so I shall argue.

Another respect in which the present interpretation is indebted to Kochen's 1978 ideas concerns the explanation of EPR-type correlations. A key feature of Kochen's account was to allow each of the component systems to have certain definite dynamical properties after their interaction, while at the same time denying that a measurement on one system can change the dynamical properties of the other (thus respecting what in Chapter 1 I called Einstein locality). The present interactive interpretation shares this feature of Kochen's 1978 views. But once more there are also important differences. In the present interpretation, irreducible dynamical properties of the compound system play a crucial role in the explanation of the correlations: But Kochen (1978) does not discuss the dynamical properties of the compound system, and does not introduce the idea of an irreducible property. Moreover, it is a significant advantage of the present interpretation over Kochen (1978) that its account of EPR-type correlations has a natural Lorentz invariant extension.

After this brief discussion of how the present interactive interpretation incorporates and transforms two of the more important ideas of Kochen (1978), I shall now consider a number of objections that have been made against Kochen (1978) in order to show how the present interactive interpretation is able to overcome each objection.

Kochen (1978) considered only interactions between two quantum systems, assumed to be otherwise isolated. But the real world presumably consists of a huge number of quantum systems, many of which are now interacting simultaneously, and many more of which have interacted in the past. It seems then that Kochen (1978) has little or nothing to say about the real world, and so can contribute little or nothing to an understanding of how quantum mechanics applies to it. A crucial respect in which the present interactive interpretation generalizes Kochen (1978) is that it applies immediately to multiple simultaneous interactions. Irrespective of its internal structure, and no matter how complex the interactions among it components, any quantum system as a whole either is or is not interacting in some way with its environment. In the present interpretation, the fundamental probabilities in quantum mechanics concern the correlated dynamical states of an arbitrary system and its environment when the two are not interacting. Multiple simultaneous interactions between one system and several others may be viewed as just one complex interaction between that system and its environment. Quantum mechanics is then concerned with the probabilities of the possible outcomes of this interaction. More technically, there is not just a single biorthogonal expansion of a vector in the universal system representative, but rather a biorthogonal expansion corresponding to each of its nontrivial subsystems. Through the subspace decomposition condition, each such expansion comes to play a role in constraining the dynamical properties of the systems concerned, provided they are no longer interacting with one another, independently of whether or not their subsystems continue to interact.

The first objection I raised in Healey (1981) concerned the time at which an interaction is concluded. This was a critical issue for Kochen (1978), since in his view a system's dynamical properties are realized only at this time. I presented Kochen (1978) with a dilemma. Either one assumes that an interaction is completed only when the interaction terms in the total Hamiltonian are strictly zero, in which case it becomes doubtful whether this ever occurs, or one assumes that an interaction is completed when these interaction terms fall below a certain limit, in which case there seems to be no nonarbitrary way of setting this limit.

This dilemma is resolved as follows in the present interactive interpretation. Note first that it is not only at the completion of an interaction that a system possesses dynamical properties. In the present interpretation a system has a dynamical state at all times, whether or not it is interacting with another system. During an interaction, the dynamical state may undergo a sequence of indeterministic transitions; but as the strength of the interaction decreases, so do both the probability of any further transition, and also the probability that the present dynamical state is very different from the asymptotic free dynamical state. In this sense, the dynamical state of a system "settles down" at or near its theoretical final dynamical state as the interaction progressively decreases. Though the fundamental probabilities in quantum mechanics do indeed directly concern the dynamical state of a system only when it is free of interaction (since the relevant interaction terms in the total Hamiltonian are zero), as the strength of an interaction decreases it becomes increasingly likely that the subspace decomposition condition will apply (at least approximately), and that the quantum mechanical probabilities of the actual dynamical state will hold. This explains why it is legitimate to apply quantum mechanics to a system even when that system is never strictly free from external interaction.

A second objection raised in Healey (1981) concerned the structure of the correlated sets of dynamical properties realized at the conclusion of an interaction. Kochen (1978) took it that each set has the structure of a Boolean σ-algebra: At least typically this algebra is generated by a set of projection operators – namely, the projections onto the subspaces M_k^σ ($M_k^{\bar\sigma}$) from the privileged set for σ ($\bar\sigma$) specified in the subspace decomposition condition of Chapter 2. If the decomposition whose existence is guaranteed by the biorthogonal decomposition lemma is unique, then each of these subspaces will be one-dimensional. Each Boolean σ-algebra they generate will then be maximal, and the dynamical state may be thought to derive from the assignment of a precise real value to some maximal quantity (that is, a dynamical variable represented by a maximal self-adjoint operator). But if the biorthogonal decomposition is *not* unique, then the Boolean σ-algebras generated by it will not be maximal, and the corresponding dynamical state would not include the assignment of a precise value to any maximal quantity. In Healey (1981) I took it to be Kochen's (1978) view that each dynamical state does in fact include a precise value assignment to some maximal quantity (so that each Boolean σ-algebra is in fact maximal), and pointed out that he had not shown how the details of the interaction specify a unique maximal extension of a non-maximal Boolean σ-algebra generated by a nonunique biorthogonal decomposition.

In the present interactive interpretation, it is not assumed that each dynamical state includes the assignment of a precise real value to some maximal quantity, and, consequently, no analogous problem arises. The conditions on dynamical states yield a unique dynamical state for a noninteracting system, even though its system representative in that state may not be a one-dimensional subspace. It is true that there is a related problem concerning an exceptional class of measurement interactions in which the apparatus system's dynamical state fails to include a precise value for the pointer position quantity, since the apparatus system representative spans vectors that intuitively correspond to distinct pointer positions. But this problem was solved in Chapter 3 by showing that such exceptional apparatus system dynamical states would be metastable even if they were to occur: Environmental interactions would rapidly modify the apparatus dynamical state so that the pointer position quantity acquired a definite value.

In his 1978 manuscript, Kochen imposed a kind of stability condition on dynamical properties, intended to show how a system may continue to possess such a property through multiple interactions. According to this condition, if a system has a dynamical property following one interaction, and the Boolean σ-algebra of properties for a second interaction also includes that property, then the system continues to have that property after the second interaction. In my (1981) comments

222

I criticized this condition on two grounds. The first criticism was that there seemed to be many interactions that violated this condition: For example, any scattering interaction in which an incoming particle with definite momentum interacts then separates with definite, but different momentum. The second criticism was that imposing this condition makes it very difficult, if not impossible, to supplement Kochen's (1978) ideas with an adequate account of measurement interactions, since a typical measurement interaction will be one in which an apparatus system initially has a property (of having a definite value for the pointer position quantity) which appears also in the Boolean σ-algebra corresponding to the measurement interaction. But to require that the apparatus system continue to have that property after the interaction is to prevent that system from recording any nontrivial result of the measurement.

Accordingly, the present interactive interpretation does not incorporate Kochen's (1978) stability condition. It does, however, incorporate a related condition which plays an important role in explaining how a system's properties may remain stable through interactions, and also how it is that measurement interactions may come to mimic the effects of the projection postulate. The statement and application of this stability condition is given in Chapter 3. The essential difference from Kochen's (1978) condition is that although appearing in the Boolean σ-algebra for an interaction remains a necessary condition, it is no longer a sufficient condition for a property to be stable through that interaction. A sufficient condition for stability is not met either in a measurement-type interaction or in a variety of other kinds of interaction (such as scattering, which produces a change in momentum). Consequently, the objections to Kochen's (1978) stability condition cannot be raised against the analogous condition imposed within the present interpretation.

Perhaps the most fundamental objection to Kochen (1978) was briefly sketched in Healey (1981) and developed more fully by Norman (1981). Kochen (1978) maintained that a quantum system has both an individual state and a statistical state. The individual state may be identified with what I call its dynamical state: It specifies what dynamical properties the system has. The statistical state specifies the probability that any given dynamical property would be realized in an appropriate interaction.[16] For Kochen (1978), the basic probabilities in quantum mechanics are those given by the Born rules, and these concern the chance of a given dynamical property's being realized in an appropriate interaction. An adequate epistemology for quantum mechanics therefore requires that there be some way of observing the individual state of a system to see what dynamical properties have in fact been realized in an interaction. The objection is that given Kochen's (1978) assumptions, there can be no adequate epistemology for quantum mechanics.

The argument may be illustrated by the example of the modified Bohm version of the EPR thought-experiment described in Chapter 4. After the initial interaction between the two spin-½ particles, the individual state of each includes either the dynamical property *spin-up in the z-direction* or the dynamical property *spin-down in the z-direction* – and one particle has each property. If these individual states are accessible to observation, it must be possible to interact appropriately with either particle in such a way that the relevant property is revealed. And this condition can initially be met, since each particle is also associated with a statistical state which determines the probability that any dynamical property will be revealed in an appropriate interaction. If particle A has the property *spin-up in the z-direction* this will be revealed in an appropriate interaction, provided that its associated statistical

[16]It is unclear whether Kochen (1978) intended to restrict the class of appropriate interactions to measurement-type interactions, and, if so, how he intended to delimit that class. The present objection requires only that those interactions which were called *M*-suitable in Chapter 3 be appropriate. If *all* interactions are appropriate, then additional objections arise.

224

state corresponds to an eigenstate of z-spin with positive eigenvalue. Similarly, the statistical state associated with particle B corresponds to an eigenstate of z-spin with negative eigenvalue.

If particle B is now subjected to an M_B measurement interaction, this does not change the individual state of particle A. But if the result is, say, spin-up, then (as shown in Chapter 4) it is certain that an M_A measurement interaction with A will also give result spin-up. Since Kochen (1978) assumes that A has a statistical state that specifies the probability for realization of an arbitrary dynamical property in an appropriate interaction, A must have such a statistical state after the initial M_B interaction. Indeed, this must correspond to the state ψ_+ of Chapter 4, since only this will yield probabilities consistent with those derived from the joint quantum state by conditionalizing on the result of the M_B measurement. Hence, the M_B interaction has changed the statistical state of A, while leaving its individual state unaffected. But now the individual state of A has become a "ghost state," inaccessible to further observation. For the only way to observe this state would be through an interaction with A appropriate for a measurement of z-spin. But it is not the individual state of A but its statistical state which determines the probabilities for the possible outcomes of this interaction, and it does so in such a way that the outcome is by no means certain to be spin-up. The interaction with B has rendered the individual state of A causally irrelevant to the outcome of future interactions involving A, and hence inaccessible to observation.

Generalizing from this example, it follows from the assumptions of Kochen (1978) that interactions with a system β may render the individual state of another system α which has previously interacted with β observationally inaccessible. If α is an apparatus system which has interacted with an object system β, then further interactions with β may render the recording of the result on α observationally inaccessible.

Whereas if the roles of α and β are reversed, further interaction with an apparatus system may render the value of the measured quantity on the object system observationally inaccessible.

Notice that the argument underlying this objection required Kochen's (1978) assumption that a quantum system has a single statistical state which correctly determines probabilities for the possible outcomes of an arbitrary appropriate interaction that system may undergo. This assumption is closely related to the common assumption that a system always has a unique quantum state. Since the present interpretation rejects the latter assumption, it is not surprising that it also rejects the former. In the present interactive interpretation, this provides the simple answer to the above objection. But I shall not rest with the simple answer since an examination of exactly how the present interpretation avoids the objection will help illuminate certain important features of the interpretation.

Consider once more the modified Bohm version of the EPR thought-experiment analyzed in Chapter 4. In the present interpretation, the A systems divide initially into two disjoint subsets: A fraction, of approximately $\cos^2\theta$, have z-spin-up, while the others have z-spin-down. Call the former subset E_1. Following the M_B interaction, the A systems may be divided into two different disjoint subsets: A fraction, of approximately $|N_+|^2$, correspond to B systems which gave result φ-spin-up, while the others correspond to B systems which gave result φ-spin-down. Call the former subset E_2. One can tell whether a particular A system belongs to E_2 by noting the result of the φ-spin measurement on its corresponding B system. Each system in E_2 will give result ψ-spin-up in a subsequent M_A measurement interaction.

Moreover, even after the M_B interaction, each A system in E_1 continues to have z-spin-up. And at any time before the M_A interaction it would have been possible to observe whether or not a particular A system is in E_1 by means of an

M-type interaction involving it that was suitable for z-spin. In this way, the dynamical property z-spin-up was still observable even after the M_B interaction: The hypothetical z-spin measurement would have faithfully revealed its presence on just those systems in E_1. Hence, any system A_i in $E_1 \cap E_2$ satisfies two conditions after the M_B interaction but before the M_A interaction: (C_1) A_i has z-spin-up, and (C_2) A_i would give result ψ-spin-up in a subsequent M_A measurement interaction. Moreover, if, instead of the M_A interaction, an interaction had been performed on the A systems suitable for a measurement of z-spin, then it would have been possible to determine whether or not a given A system was in $E_1 \cap E_2$ by noting the result and comparing it with the result of the M_B interaction on the corresponding B system.

Consider the state of an element A_i of $E_1 \cap E_2$ in the interval between the M_B and M_A interactions. A_i is an element of E_2, and may also be assigned a statistical state (corresponding to ψ_+) which will yield the correct probability for observation of any dynamical property of A_i in an appropriate subsequent M-type interaction, given condition C_2 alone. Moreover, there are ways to observe that C_2 holds (e.g., by performing the M_B interaction on the corresponding B system and using the fact that if the result is ϕ-spin-up, the result of the M_A interaction will certainly be ψ-spin-up), and these observations form part of a preparation procedure for ψ_+ which has actually occurred. A_i is also an element of E_1, and may in other circumstances have been assigned a statistical state (corresponding to ζ_+) which would yield the correct probability for observation of any dynamical property of A_i in an appropriate subsequent M-type interaction, given condition C_1 alone. But although one can easily describe preparation procedures for ζ_+, these have not in fact occurred. If, instead of the M_B interaction, an interaction had been performed on the B systems appropriate for a measurement of z-spin, then it would have been possible to observe whether or not A_i was in E_1 by noting the result for its corresponding B system, and using the anticorrelation in z-spin values between corresponding A and B systems; and these observations would then have formed part of a preparation procedure for ζ_+. However, only if some such preparation procedure had actually occurred would it have been legitimate to ascribe statistical state ζ_+ to A_i.

Consequently, A_i may be assigned either statistical state ψ_+ or statistical state ζ_+, depending not on its own dynamical state, but rather on the nature and results of interactions involving its correlated B system. But there is no single statistical state which may be assigned to A_i by virtue of its membership in $E_1 \cap E_2$ which will yield correct probabilities for observation of dynamical properties of A_i in an appropriate subsequent M-type interaction, given the compound condition $C_1 \& C_2$. And this accords with the absence of any preparation procedure for $E_1 \cap E_2$, and with the impossibility of observing prior to the M_A interaction that the condition $C_1 \& C_2$ holds on A_i in a way that does not disturb the probabilities associated with the possible outcomes of the M_A interaction. No quantum or statistical state may be ascribed to an A system independent of the specification of any preparation procedure and subsequent measurement interaction. And since the z-spin and ϕ-spin interactions are mutually physically exclusive, it is physically impossible to specify a preparation procedure for $E_1 \cap E_2$. This is the physical reason why there can be no single statistical state of A_i corresponding to $E_1 \cap E_2$.

A system A_i in $E_1 \cap E_2$ meets conditions C_1 and C_2. Moreover, A_i is certain to reveal z-spin-up in an appropriate M-type interaction, even though this does not follow from the assignment to A_i of the statistical state corresponding to ψ_+ which is associated with the set E_2. In this way the dynamical state of A_i is not a ghost state, but would have immediate causal relevance to the outcome of an M-type interaction appropriate for its observation. And this illustrates how it is that in the present interpretation the dynamical state of a system remains observationally accessible, even after an interaction with a correlated system which in Kochen's (1978) view would have rendered it observationally inaccessible.

A final objection to Kochen (1978) concerned his account of the origin of EPR-type correlations, and, in particular, of the perfect correlation between a positive result for a measurement of ϕ-spin on a B system and of ψ-spin on its correlated A system. After a measurement of ϕ-spin on B_i has given a positive result, a measurement of ψ-spin on A_i is also certain to give a positive result. The objection concerns Kochen's (1978) explanation of why this is so. For there is no feature of the individual (dynamical) state of A_i following the M_B measurement interaction which grounds this certainty: In particular, A_i does not have ψ-spin-up prior to the M_A interaction. Kochen (1978) therefore appeals to the statistical state of A_i as grounding this certainty. Now he stresses that the statistical state is epistemological rather than ontological: It represents our knowledge of a system rather than the actual condition of the system. And this is important for Kochen (1978), since he acknowledges that the statistical state of A_i is altered by the M_B interaction, but argues that this does not represent any physically problematic action-at-a-distance, since the change merely takes account of the additional information made available following the M_B interaction. But if the statistical state of A_i has merely epistemological significance, then it cannot itself serve to ground the certainty of the result of the ψ-spin measurement on A_i: It can at most express that certainty. Kochen (1978) cannot explain how any actual property of any actual system serves to ground this puzzling certainty, and he even denies that any property of A_i represented in its individual state does so.

In the present interactive interpretation, the certainty is grounded not in any property of A_i itself, but rather in properties of the compound system $A_i \oplus B_i$, including the key irreducible correlational property discussed in Chapters 4 and 5. It is by virtue of its holism that the present interpretation is able to adhere in this situation to the realist principle that any disposition expressed by a true subjunctive conditional requires a categorical basis. And it is because the process it postulates to account for this and other EPR-type correlations is nonseparable that the present interpretation is able to conform to all defensible requirements of both relativistic invariance and locality, as explained in Chapter 5.

While Kochen's 1978 views were unpublished, a paper (Kochen, 1985) has since appeared in which he presents a later version of these views. Although Kochen (1985) had no influence on the development of the present interactive interpretation, it remains of some interest to compare the two.

There are two main respects in which Kochen (1985) differs from the earlier, unpublished Kochen (1978). Just like the present interactive interpretation (though apparently independently of it) Kochen has come to generalize his interpretation so that it applies to more then just two otherwise isolated systems. He, also, now treats interactions as fundamentally occurring between a system and its entire environment, and then applies the polar decomposition theorem to determine the states of the interacting systems. In this respect Kochen's views have come closer to the present interactive interpretation, and the parallel independent evolution of our views is to be welcomed as a possible indication that between them they contain elements of the truth.

But Kochen (1985) also contains the novel notion of a **witnessed state,** whose application is problematic and may well represent a significant divergence from the basic ideas of the present interactive interpretation. For Kochen (1985), the polar decomposition of the compound state of a system and its environment defines *at any moment* the state of each, as witnessed by the other.[17] The witnessed state of a system

---

[17] Notice that this already represents a significant difference from the present interactive interpretation. In the present interpretation, though a system is assumed to have a dynamical state at any moment, it is only when a system is not interacting that the system representative condition is assumed to constrain this (cf. Chapter

is intended to play much the same role as Kochen's (1978) individual state: It indicates what is true of the system at that moment. Mathematically, the witnessed state of a system may be regarded either as what I have called its system representative, or as a truth-value assignment to a subset of its dynamical properties generated by this. The subset consists in the Boolean σ-algebra of properties generated by projections onto the accessible subspaces for the biorthogonal decomposition of the instantaneous compound state vector. And the system has a property under this truth-assignment if and only if the system representative is included in the subspace corresponding to that property. But although mathematically well defined, the significance of the witnessed state is quite unclear.

There is a basic ambiguity in the notion of a witnessed state: Does this represent the actual dynamical state of a system, independent of any system which may in fact be interacting with it, or does it represent only the dynamical state *from the perspective of a particular interacting system*? For the first alternative, the witnessed state of a system is just its dynamical state. But this choice has unfortunate consequences. In the example developed in Chapter 4 [and considered also by Kochen (1985)], the system representative of $A \oplus B$ after the M_B interaction is (say) $[\psi_+^A \otimes \phi_+^B]$, whereas that of A is (say) $[\zeta_+^A]$. It follows that A has the witnessed property z-spin-up (corresponding to the subspace spanned by ζ_+), and $A \oplus B$ *lacks* the property that corresponds to the subspace projected onto by $\mathbf{P}_{\zeta_+} \otimes \mathbf{I}$: whereas $A \oplus B$ has the witnessed property corresponding to the subspace projected onto by $\mathbf{P}_{\psi_+} \otimes \mathbf{I}$ but A lacks the property ψ-spin-up! But this contradicts a basic tenet of the quantum mechanics of interacting systems: that the operator $\mathbf{A} \otimes \mathbf{I}$ on $A \oplus B$ corresponds to the

2). It is hard to see how Kochen's view could be supplemented by a coherent stochastic dynamics for witnessed states if they are *always* constrained in accordance with the polar decomposition theorem, as Kochen (1985) maintains.

231

quantity $\mathcal{A}$ on A. Moreover, any plausible account of measurement interactions will imply that an interaction suitable for measuring $\mathcal{A}$ on A is also suitable for measuring the quantity corresponding to $\mathbf{A} \otimes \mathbf{I}$ on $A \oplus B$, and must give the same result. These constitute powerful reasons for rejecting the first way of resolving the ambiguity in the notion of a witnessed state. Notice that it is because the connection between a system's dynamical state and its system representative is less tight in the present interpretation (see the system representative condition of Chapter 2) that the present interpretation does not face an analogous difficulty.

If Kochen's (1985) notion of a witnessed state is understood perspectivally, then this difficulty does not arise. For then the state of A as witnessed by $B \oplus M_B$ may include the witnessed property z-spin-up but not the property ψ-spin-up, whereas the state of $A \oplus B$ as witnessed by M_B simultaneously includes a property corresponding to $\mathbf{P}_{z_+} \otimes \mathbf{I}$ but not a property corresponding to $\mathbf{P}_{z_+} \otimes \mathbf{I}$. The point is that a system has no absolute dynamical state, but only dynamical states as witnessed by various interacting systems: And there is no reason why these must always coincide. But this solution to the difficulty involves a radical break both with Kochen (1978) and with the present interactive interpretation. For each of these maintains that a system has a single, nonperspectival, dynamical state, and interprets the quantum mechanical probabilities to concern the probability that a system will assume one rather than another such state in appropriate circumstances. If Kochen (1985) is taken to deny this basic assumption, then this brings his view much closer to Everett's, and raises serious interpretative difficulties, including the following: If the quantum mechanical probabilities concern the witnessed properties of a system, then how are these related to the system's *actual* properties (if any)? How do a system's various witnessed states hang together? And, especially, why is it that our observations, including quantum measurements, all seem

232

consistent with the natural view that the world consists of systems which may be in various (absolute) states, irrespective of how these are witnessed (if at all)?

Neither Kochen (1978) nor Kochen (1985) contains adequate accounts of quantum state preparation or of the relation between quantum states and dynamical states ['individual states,' in the terminology of Kochen (1978), or 'witnessed states,' in the terminology of Kochen (1985)]. This is an incompleteness which the present interactive interpretation seeks to correct. Though Kochen assumes that the system + environment compound has a quantum state, it is also necessary for him to justify the successful practice of ascribing quantum states to subsystems of this compound. He acknowledges that for his view this practice cannot be justified if the quantum state of a system is simply identified with its dynamical state, for even when the latter may be represented by a state vector, this does not always yield the correct probabilities when the Born rules are applied to it [cf. the above discussion of Kochen's (1978) treatment of the EPR-type correlations]. But then it is necessary to explain under what circumstances it is legitimate to ascribe a quantum state to a subsystem, how this quantum state is determined, and how quantum states are related to dynamical states.

One difficulty here concerns the projection postulate. As with the present interactive interpretation, Kochen (1985) takes this to correspond not to any physical process, but rather to a change in perspective. But his 1985 paper offers no general account of when such a change is appropriate. Moreover, any interpretation which denies that projection is a physical process owes an explanation of those phenomena which are so readily explained by the assumption that it is – especially the possibility of quantum state preparation and the verifiability of measurement results. But Kochen offers no explanation of either of these phenomena. Taking its cue from Everett's interpretation, the present interactive interpretation offers explanations of both phenomena. The verifiability of measurement results is explained by appeal to the stability condition on dynamical states. Quantum state preparation is explained by showing how it is that the basic probability rules (derived from the subspace decomposition condition) imply that the Born rules hold for systems which have undergone suitable *P*-type interactions and which will undergo suitable *M*-type interactions, provided that one ascribes to such systems exactly the quantum state which the *P*-type interaction may be taken to prepare. Both these explanations were given in Chapter 3. By explaining why, and in what circumstances, it is legitimate to ascribe a quantum state to a system, the present interactive interpretation is able to offer a coherent account of the character of quantum states and their relation both to dynamical states and to an observer's knowledge of these. That account makes it clear how it is that the quantum state of a system has both ontological and epistemological aspects, representing neither the dynamical state of the system nor simply an observer's knowledge of it, but rather a summary of information about objective probabilistic dispositions concerning the outcomes of future interactions involving that system – information of just the kind which is likely to be both available to an observer and useful to him in predicting the statistics of measurement results on ensembles of quantum systems.

7

Open questions

The interactive interpretation I have presented here leaves many interesting questions unresolved. In conclusion, I shall raise several such questions, and discuss their significance for the interpretation. These may be divided roughly into two kinds: Some are merely requests for a more complete development of the view expressed here, whereas others represent unresolved challenges to its adequacy as an interpretation of quantum mechanics. What these questions all have in common is that at the present time I cannot myself adequately answer them. I raise the questions here to commend them to others, as suggesting interesting topics for further research of a kind that holds out hope for an improved understanding of the quantum world, as well as of our current best theories of it.

The first such question concerns the dynamics of quantum systems during interactions. The present interactive interpretation rests on certain rather weak assumptions about how the dynamical state of a quantum system changes during an interaction: It would clearly be of great interest to explore further details of this process. Such an exploration requires both theoretical and experimental research. If the present interpretation of quantum mechanics is along the right lines, then not only does a quantum system have a well-defined dynamical state at each moment, but also there is an interesting stochastic element to the evolution of that dynamical state during an interaction. Although the present interpretation constrains this stochastic behavior, it neither specifies nor determines it. The following constraints have been mentioned earlier (in Chapter 2). The major constraint is imposed by requiring that the cumulative effect of changes in the dynamical state during an interaction is such that when the interaction is over the subspace decomposition condition holds, in conjunction with the system representative condition. Subsidiary constraints are imposed by the stability condition, and by the qualitative requirement that the magnitude and frequency of changes of dynamical state decrease with decreasing interaction strength. The obvious question then arises as to what kind of dynamics can meet these constraints, prescribing accurate laws for the evolution of the dynamical state of a quantum system throughout an interaction.

This is a question to be approached both theoretically and experimentally. Theoretically, what is required is the creation and investigation of detailed mathematical models of relevant stochastic processes to see what kinds of models can meet all necessary constraints. Experimentally, it will be necessary to devise experiments capable of discriminating between the predictions of a number of such models concerning the dynamical state of a quantum system while it is subject to significant interaction. There are two reasons to expect that such experiments may prove very difficult to conduct. First, it seems that they would need to be performed within very short time intervals, in order to investigate the nonequilibrium behavior of quantum systems (during interactions). Second, such experiments might potentially reveal effects that are not predicted by the application of quantum mechanics, as this theory is usually understood, to such short-lived and unstable configurations – but no experiments yet performed have presented quantum mechanics with any such challenge.

In lieu of a concrete mathematical model, I can offer only a pictorial analogy for the evolution of dynamical states during an interaction. Assume that the system representative condition holds, but the subspace decomposition condition fails, during an interaction. Take the evolution of the system representative during an interaction to be composed of a discontinuous indeterministic process superimposed on a continuous deterministic process. Think of the resulting motion as an abstract analog in Hilbert space to the motion of a bean bag in the interior of a cylinder rotating about its horizontal axis in ordinary Euclidean space. The rotation of the cylinder corresponds to the rotation in Hilbert space of the basis of accessible subspaces picked out by the biorthogonal decomposition lemma. For much of the time, the system representative evolves smoothly, perhaps in accordance with a generalized Schrödinger equation governed by the noninteracting Hamiltonian for the system. This continuous aspect of the motion corresponds to the bean bag's remaining in contact with the cylinder and so being lifted up its side as the cylinder rotates. But as the interaction proceeds (the cylinder rotates further) the system representative is carried by this continuous evolution further and further from a "stable" subspace – one which is accessible according to the subspace decomposition condition – and it becomes more and more likely that the system representative will undergo a discontinuous and indeterministic transition to an instantaneously accessible subspace distant from it in the Hilbert space metric (just as the bean bag becomes more and more likely to fall from the side of the rotating cylinder onto its instantaneous "floor"). This analogy is not close. For example, although it may allow for a discontinuous transition to occur at random, it does not seem to permit a random selection from a number of possible states into which this transition may occur. Moreover, nothing in the analogy seems to correspond to the varying strength of the interaction (though perhaps this might be taken to be the diameter and/or the speed of rotation of the cylinder). It is another open question as to whether or not this analogy will prove fruitful in suggesting detailed models for the evolution of dynamical states during interactions.

It is important to note that acceptance of the present interactive interpretation need not await resolution of these interesting further questions concerning the detailed evolution of dynamical states during an interaction. The interpretation requires only that it be possible to postulate *some* stochastic dynamics consistent with the aforementioned constraints, not that these dynamics be experimentally verified or even theoretically natural. Only if *no* possible dynamics could meet the constraints would this constitute a reason to doubt the interpretation. For though the present interpretation may indeed prove fruitful in the discovery of new physics, it is not appropriate to judge its adequacy by such success. The interpretation is offered neither as a new theory nor as a sketch of a new theory, but rather as a way of understanding a familiar though puzzling theory, namely, nonrelativistic quantum mechanics.

This last point connects with a second question: How far does the present interactive interpretation lend itself to generalization to relativistic quantum theories? In responding to this question it is important to distinguish between particle theories and field theories. Although the extension to an interpretation of the relativistic quantum theory of a fixed array of particles such as the Dirac theory for fermions or the Klein-Gordon theory for bosons seems relatively unproblematic, the extension to a relativistic quantum field theory in which "particles" are treated as excitations of an underlying field, and particle number is represented not by a constant but by a self-adjoint operator, appears fraught with difficulties.

Modifications of three kinds will be necessary to render quantum mechanics, according to the present interpretation, a Lorentz invariant theory. The kinematics of the theory must be formulated so that the Lorentz transformation properties of both quantum states and dynamical states are manifested; equations of motion must be reformulated so that they are Lorentz covariant and pick out no preferred frame; and all

238

specifications of states at a time must be replaced by specifications of states on a spacelike hyperplane (or more general
spacelike hypersurface).[1]

Perhaps the simplest way to formulate the kinematics
would begin with the system representative. The Lorentz
transformation properties of this could be made manifest in
basically the same way that one specifies the Lorentz transformation properties of a relativistic quantum state (e.g., the
wave-function of the Dirac electron): Each system representative would transform in accordance with an appropriate
representation of the inhomogeneous Lorentz group. The
Lorentz transformation properties of dynamical states and of
quantum states would then be determined in accordance with
relativistic generalizations of the conditions given in Chapter 2.

In order to transform the time-dependent Schrödinger
equation into the Dirac or Klein-Gordon equation, one mod-
ifies Chapter 2's law of free evolution so that the process it
describes is Lorentz invariant and described by a precursor
of one of these two equations. The law thus formulated cannot itself be identified with the Dirac (or Klein-Gordon, respectively) equation, for these latter equations describe the
evolution of quantum states, not dynamical states. Rather,
the usual relativistic laws of evolution for quantum states
derive from the closely related relativistic laws of evolution
for underlying dynamical states, just as in the nonrelativistic
case.

The stability condition of Chapter 2 needs only minor reformulation to meet the constraint of Lorentz invariance. It
is necessary merely to replace the terms 'before' and 'after'

---

[1]In order to render the interpretation Lorentz invariant, it is only necessary to
consider state assignments on spacelike hyperplanes. It may seem more general and
elegant to formulate a Lorentz invariant generalization of the present interpretation
so that it assigns states on arbitrary spacelike hypersurfaces. However, Gordon
Fleming has suggested to me that it is very difficult to formulate a *local,* Lorentz
invariant, relativistic quantum theory in this manner. Clearly, any further generalization, to a curved relativistic space-time, would require such a formulation.

as they occur in that condition, either explicitly or implicitly (through the arrow which represents the transformation brought about by the interaction), by 'intersecting (only) the causal past of' and 'intersecting (only) the causal future of,' respectively. This presupposes that the interaction is confined to a certain region of space–time, and that attributions of system representatives are made on spacelike hyperplanes which bear the appropriate geometric relation to this region. The relativistic reformulation of the stability condition places no constraints on the system representative on any spacelike hyperplane which intersects the interaction region, just as the original stability condition places no restrictions on the system representative at any time during the interaction. Note that the relativistic generalization of the free evolution condition also implicitly assumes that the evolution it describes characterizes the dynamical state of a system only when this is defined on a spacelike hyperplane in such a way that its location on that hyperplane is restricted to a region which does not intersect the space–time region occupied by any interaction.

The subspace decomposition condition must also be reformulated in accordance with the requirements of Lorentz invariance. This is readily done when a system representative is assigned on a spacelike hyperplane by replacing the antecedent condition – that a system not interact at a time – by the condition that the region to which its position is confined, on the spacelike hyperplane on which the system representative is defined, be one which does not intersect a space–time region occupied by any interaction (since the universal Hamiltonian contains no nonvanishing interaction terms defined at any space–time point within that region).

Such natural reformulations seem all that is required to generalize the present interpretation to an interpretation of a relativistic quantum theory in which the quantum state describes the likely observed behavior of a determinate and fixed number of particles. But it is well known that a relativistic theory of this kind is at best a stepping stone on the way to a relativistic quantum field theory. Field theories are required to describe systems (such as coherent light) which do not consist of a determinate, fixed, set of particles. Moreover, the conceptual difficulties of relativistic c-number theories (such as those associated with Dirac's sea of negative energy electrons) receive an adequate resolution only in a relativistic quantum field theory (if at all). Can the present interactive interpretation be generalized to an interpretation of a relativistic quantum field theory?

The present interactive interpretation has a number of features which indicate that any such generalization will require extensive modification, at least of the technical framework presented here. The hierarchy of quantum systems described in Chapter 2 presupposed the existence of a fixed set of atomic quantum systems. But this presupposition clearly seems false in a quantum field theory. If one takes the quanta of the field to be atomic systems, then their number is not fixed and may not even be definite. If, on the other hand, one takes the space-time points, or regions, at which the field is defined as atomic quantum systems, then these constitute a nondenumerably infinite set. Moreover, the subspace decomposition condition presupposes the existence of a complement $\bar{\sigma}$ for each quantum system σ, and in Chapter 2 this was taken to satisfy $\sigma \oplus \bar{\sigma} = \Omega$, where Ω is the universal system. But if there is no fixed finite set of quantum systems, it is not clear what could be meant by Ω, and, therefore, by $\bar{\sigma}$. Consequently, there seems to be no obvious way of generalizing the subspace decomposition condition of the present interactive interpretation to the domain of relativistic quantum field theories.

Even though the technical framework presented here requires extensive overhaul to suit it to the domain of relativistic quantum field theories, the basic ideas of the present interactive interpretation may well prove important in interpreting quantum field theory. I expect this to be true in particular of the following central ideas. First, there is the idea that

241

measurement may be understood as a physical interaction of a certain kind, whereas quantum theory itself may be used to say what kind of interaction that is. Then, there is the idea that the quantum state of a system must be conceptually distinguished from its dynamical state, and that when this is done the measurement problem is readily soluble. And, finally, there is the idea that holism and nonseparability, understood quite precisely, hold the key to understanding the explanation quantum theory gives of many physical phenomena, including phenomena (such as violation of the Bell inequalities) which seem to threaten cherished physical and/or metaphysical principles about local action.

How important is it that the present interactive interpretation have a natural generalization to the relativistic domain? Although it is important that it generalize readily to a relativistic c-number theory, it seems much less important that it generalize in a straightforward manner to relativistic quantum field theories. Only by exploiting features of the present interpretation in the context of a relativistic space-time was it possible to offer a convincing account of violation of the Bell inequalities in the present interpretation (see Chapter 5). If the present interpretation had no natural generalization to that setting, then the interpretation would remain seriously inadequate. On the other hand, I do not take it to be a major defect of the present interpretation that it possesses no simple generalization to the very different context of a relativistic quantum field theory. For there are reasons to believe that such a theory is conceptually quite distinct from nonrelativistic quantum mechanics in exactly those respects which are most important to nonrelativistic quantum mechanics, for the present interpretation. The basic ontology of a quantum field theory is not yet clear (fields, particles, or space-time points?), but whatever it is, it is likely to be in some respects discontinuous with that of nonrelativistic quantum mechanics. Central field-theoretic concepts such as renormalization have not yet been given a wholly unproblematic interpretation, and failing such an interpretation quantum field theory itself is much less well understood than nonrelativistic quantum mechanics, providing that understanding is likely to prove an extraordinarily difficult task. And it may well turn out that the best interpretation of relativistic quantum field theory will be very different in central respects from the best interpretation of nonrelativistic quantum mechanics. In light of these points it would clearly be unreasonable to criticize the present interpretation for not generalizing naturally and easily to give an acceptable interpretation of relativistic quantum field theory.

Though the questions raised so far in this chapter suggest interesting and potentially fruitful topics for future investigation, failure to arrive at satisfactory answers would not, in my opinion, substantially detract from the worth of the present interactive interpretation. But there are other open questions, whose answer is a more pressing matter. The first such question has to do with the treatment of measurement and state-preparation interactions.

Recall that Chapter 3 offered no precise definitions of the entire class of M-type or P-type interactions, but rather, concentrated on exhibiting simple and rather idealized elements of these classes in order to establish that such interactions are at least possible. The exact scope of these classes remains an open question. Moreover, it appears critical to the success of the present interactive interpretation to exhibit a number of *actual* measurement and preparation procedures which (at least approximately) fall within these classes. For quantum mechanics (including the Born rules) is applied in circumstances where some actual interaction is taken to constitute or permit a measurement of some quantum mechanical dynamical variable on systems which have been subjected to some definite prior interactions. And such an application can only be finally justified under the present interpretation if it is possible to argue convincingly that the interactions involved are (respectively) M-type and P-type interactions, or at least that each is a close enough approximation to such an interaction for it to be legitimate to apply the Born rules within the experimental errors involved in the given application.

What is required, then, is a detailed analysis of particular experimental procedures on the basis of the present interpretation. This will suggest alternative ways of modeling actual measurements and preparations in terms of M-type and P-type interactions, and, hence, will help to guide a theoretical investigation of the scope of these classes. And it will help to probe the adequacy of the present interpretation by assessing the extent to which it is possible to extend the idealized treatment of Chapter 3 to apply to more realistic interactions. Whereas §3.3 began the removal of idealizations in the treatment of M-type interactions in terms of SMIs, the application to realistic experimental arrangements will require considerably more effort. And it is clear that a faithful modeling of actual state preparation procedures will require a richer and more complex analysis of P-type interactions and their concomitants than that provided by §3.3's idealized SPPs. The analysis of actual experimental arrangements remains a major uncompleted task for the present interactive interpretation.

In this connection I wish to note the following problem, pointed out to me by David Albert. It is plausible to assume that many actual interactions involved in quantum measurements can be approximated by SMIs in a way that was not considered in Chapter 3. Consider, for example, any measurement which, as ordinarily described, involves splitting a beam of incoming particles into several beams of outgoing particles, corresponding to different measurement outcomes. A simple example of this is provided by the Stern-Gerlach experiment. It is natural to suppose that the location of a particle in one beam rather than another corresponds to a value of a recording quantity for the measurement. This is so even though it may require one or more further iterations of measurement-type interactions to amplify the result to a level at which it becomes directly perceptible. Now it is a consequence of the normal quantum mechanical treatment of the initial beam-splitting interaction that the spatial wavefunctions corresponding to different outgoing beams are never strictly orthogonal, although the overlap does decrease rapidly as the beams propagate. An initially superposed system representative will, after the initial interaction, have a decomposition which is almost, but not quite, biorthogonal when expressed in a basis of product vectors one of whose components is associated with each beam. It then follows that the actual biorthogonal decomposition does not correspond to a dynamical state in which the particles are recorded as being in a definite beam.

This situation may be modeled by an approximate SMI governed by the following interaction.

$$\chi_{ij}^{\sigma} \otimes \chi_0^{\alpha} \rightarrow \xi_{ij}^{\sigma} \otimes \theta_{ij}^{\alpha}. \tag{7.1}$$

The difference between (7.1) and (3.1) is that whereas in (3.1) we have $(\chi_{ij}^{\alpha}, \chi_{i'j'}^{\alpha}) = \delta_{ii'} \cdot \delta_{jj'}$, in (7.1) we do not have $(\theta_{ij}^{\alpha}, \theta_{i'j'}^{\alpha}) = \delta_{ii'} \cdot \delta_{jj'}$: Instead, the $\{\theta_{ij}^{\alpha}\}$ are only approximately orthogonal. It is plausible to suppose that actual measurement interactions are more faithfully modeled by (7.1) than by (3.1). But whereas the argument given in Chapter 3 may suffice to show that for an arbitrary initially superposed state of a system subjected to an interaction of the form (3.1) a result will always be recorded in the dynamical state of α, an analogous argument fails to show this for a system subjected to an interaction of the form (7.1). Can the account of M-type interactions offered in Chapter 3 be extended to cover interactions governed by (7.1)? The obstacle to such an extension is that whereas a nontrivial superposition of vectors from the right-hand side of (3.1) is already expressed as a biorthogonal expansion, a nontrivial superposition of vectors from the right-hand side of (7.1) is not. It therefore follows that, although α will always be in some definite state at the conclusion of an interaction described by (7.1), there is no reason to expect that this state will generally be one in which the recording quantity has a precise value, or even an approximately precise value.

Recall that in Chapter 3 an answer to a superficially similar problem was given which involved considering the effect of further iterated M-type interactions. The problem there concerned the measurement of dynamical variables whose associated operators do not commute with all additively conserved quantities. Since an exact measurement of such a dynamical variable is impossible, it was necessary to show how an approximate measurement might be made. But a single interaction could not be used to perform even an approximate measurement. In that case, though, it turned out that by a suitable choice of iterated M-type interactions one could indeed model a successful approximate measurement procedure. It is therefore natural to expect that the present problem is likewise amenable to solution by this method. However, there are serious obstacles to any such solution.

Suppose, for example, that one considers the effect of a second interaction between α and a further system β, immediately following an initial interaction of type (7.1). The second interaction is designed to effect a correlation between the initial dynamical state of σ and the final dynamical state of β by coupling to α after the initial approximate SMI between σ and α [represented by (7.1)]. Specifically, if this second interaction were to satisfy (7.2), then one could see how the final dynamical state of β could come to record a definite result for the measurement:

$$\theta_{ij}^{\alpha} \otimes \chi_0^{\beta} \rightarrow \theta_{ij}'^{\alpha} \otimes \chi_{ij}^{\beta}. \tag{7.2}$$

For the result of interaction (7.1), followed immediately by interaction (7.2), on an initially superposed vector would then be as follows:

$$\left(\sum_{ij} c_{ij}\chi_{ij}^{\sigma} \right) \otimes \chi_0^{\alpha} \otimes \chi_0^{\beta} \rightarrow \qquad (7.3)$$

$$\sum_{ij} c_{ij}\left(\xi_{ij}^{\sigma} \otimes \theta'^{\alpha}_{ij} \otimes \chi_{ij}^{\beta} \right).$$

And even if the $\{\theta'^{\alpha}_{ij}\}$ are not orthonormal, (7.3) will represent a biorthogonal decomposition in $(H^{\sigma} \otimes H^{\alpha}) \otimes H^{\beta}$. In that case, even though the recording quantity on α may not assume even an approximately precise value after the approximate SMI (7.1), the recording quantity on β would assume a precise value after the subsequent interaction (7.2).

Unfortunately, there can be no second interactions with the required properties, since the transformation represented by (7.2) is not unitary, and is hence disallowed by the dynamical evolution laws of quantum mechanics.

Suppose instead that one immediately follows (7.1) by an SMI suitable for measuring the value of a dynamical variable $\mathcal{A}$ on α for which the $\{\theta^{\alpha}_{ij}\}$ are approximate eigenstates. This would be the analog of performing a position measurement on a particle supposedly confined to a beam described by a wave packet centered on a particular value for the relevant spatial coordinate. The result of the iterated interactions for a nontrivially superposed state would then be a superposed state which is approximated by a biorthogonal decomposition in $(H^{\sigma} \otimes H^{\alpha}) \otimes H^{\beta}$ in which the dynamical state of β records a definite value of $\mathcal{A}$ on α. The fact that this recorded value would only very probably also reflect the value (if any) of the variable on σ which the procedure is ultimately designed to measure would then be exactly what one would expect of an approximate measurement. But in the present context, approximation to the right kind of biorthogonal decomposition is not enough to guarantee *any* definitely recorded result in β. The addition of a further M-type interaction has merely moved the original difficulty to a later stage in the measurement chain, without in any way easing that difficulty.

The problem remains of describing even an idealized model for an interaction, or sequence of interactions, capable of extracting a definite result of a measurement which begins with an initial approximate M-type interaction that satisfies (7.1). If this problem has no direct solution, then it would be necessary to give convincing arguments as to why no actual measurement interaction is in fact faithfully modeled by (7.1). For recall that if one allows that the apparently more idealized (3.1) faithfully models actual measurement interactions, then the problem we have been investigating does not arise.

Two further pressing questions concern the character of dynamical states ascribed to systems in the present interactive interpretation, and, more particularly, the behavior of these states under certain limiting operations. Chapter 1 entertained as a plausible supposition that the actual dynamical state of any macroscopic system suited to display the result of a quantum measurement is observationally indistinguishable from a definite classical state. But the truth of this supposition remains an open question. It is therefore important to investigate the extent to which it actually holds for the present interpretation. This is important because the present interpretation cannot simply assume that macroscopic M-type systems behave as if they have classical states: They are, after all, quantum systems, even though they are systems of great complexity, in continual interaction with their environment. What is required is an examination of the nature of the dynamical states of complex M-type systems, whose goal is to determine whether or not these do indeed become classical (in the sense of containing more and more sharply defined values for dynamical variables) as the complexity of the systems increases to match that of actual macroscopic recording devices; or, failing that, whether such systems come close to being observationally indistinguishable from classical systems.

A final question concerns the application of quantum mechanics in the present interpretation. Chapter 2 adopted the simplifying assumption that, in any actual application, quantum mechanics is applied to a subsystem of the universe composed of some fixed, finite set of atomic systems, and which may be taken as the universal quantum system ω for the purposes of that application. It would be theoretically possible to give a quantum mechanical description of the world under the present interpretation without making this assumption, by taking ω to be the actual universal system, composed of some fixed and definite (but not necessarily finite) set of atomic systems. But no real-life application of the theory could be based on knowledge of the universe's dynamical state, or even of its atomic composition. Any actual application must offer a simplified model of the real world by taking some restricted subsystem to constitute ω for the purposes of that application.

The following difficulty then arises. Suppose that one chooses a particular system ω_1 to constitute ω when applying the theory. An alternative choice would have been a more inclusive system $\omega_2 = \omega_1 \oplus \sigma$ (for some system σ). Based on the choice $\omega = \omega_1$, a given subsystem α of ω_1 may be assigned system representative η_1 (say); whereas, based on the choice $\omega = \omega_2$, α may be assigned system representative η_2. Even when these assignments are made in conformity with the subspace decomposition condition, it does not immediately follow that $\eta_1 = \eta_2$. But if $\eta_1 \neq \eta_2$, then it follows from the system representative condition that the dynamical state attributed to α will likely depend on which of ω_1, ω_2 one chooses to represent the universal system ω. This does not yet constitute an inconsistency within the interpretation, since neither of these choices is intended as more than a convenient approximation, made to apply the theory simply in a complex world. But it does threaten to undercut the claim that the present interactive interpretation renders quantum mechanics an explanatorily powerful theory by yielding an account of the actual dynamical processes which underlie the use of the theory to yield correct probabilities of measurement results. For if the account of these processes is quite different, depending on how one makes a relatively arbitrary choice of restricted subsystem to constitute ω when applying the theory, then it seems that the apparent explanatory power which flows from one particular choice is illusory.

In order to resolve this difficulty, what would be required would be a demonstration that application of quantum mechanics to wider and wider systems including a given system subjected to measurement interactions leads to ascriptions of dynamical states to it (and to related systems) which may plausibly be regarded as progressively more accurate approximations to their actual dynamical states. Here, too, what is required is a kind of limit result. In this case, the desired result is that, as the system chosen to constitute the universal quantum system becomes more and more complex, the dynamical states ascribed to fixed subsystems "converge" to certain limiting states – namely, the *actual* dynamical states of these systems.

Such a result is presumably demonstrable for many applications of Newtonian mechanics (given the assumption that this theory is correct). Consider, for example, applications of Newtonian mechanics to determine the trajectories of planets and other bodies in the solar system. One may calculate the orbit of (say) Mars neglecting the gravitational forces on Mars due to the other planets: Then, one may recalculate the orbit including one or more such perturbing forces, and show that the initially calculated orbit approximates to the perturbed orbit. The inclusion of further perturbing forces could then plausibly be argued to lead to still better approximations to the actual orbit of Mars.

The question that arises is "Can analogous results be established for applications of quantum mechanics in the present interactive interpretation?" This question remains an open one. Part of what would be involved in returning a positive answer to the question would be to give a clear statement of

250

the required sense of convergence. It is to be noted particularly that mere numerical convergence of predicted probabilities for measurement results would not suffice to establish the key claim – that applications based on larger and larger subsystems produce better and better approximations to the real account of the dynamical processes that underlie these probabilities.

Appendix

A set of properties ascribed in accordance with the weakening, composition, composite exclusion, prime exclusion, system representative, and minimal meshing conditions is

a. not always cogenerable, but
b. always cotenable.

A single counterexample suffices to establish part (a). Consider, then, a model universe ω composed of two atomic spin-½ systems 1,2 at a time when they are not interacting. Assume that ω is in the state

$$\psi^\omega \;=\; \cos\theta\cdot(\zeta^1_+ \otimes \zeta^2_-) \;-\; \sin\theta\cdot(\zeta^1_- \otimes \zeta^2_+) \,,$$

where $|\cos\theta|\neq|\sin\theta|\neq 0$, and $\zeta^i_\pm$ is an eigenvector of z-component of spin of system i with eigenvalue $\pm \hbar/2$ ($i=1,2$). The system representatives of $\omega,1,2$ assigned in accordance with the subspace decomposition condition are all one-dimensional, and projected onto by $\mathbf{P}_{\psi^\omega}$, $\mathbf{P}_{\zeta^1_+}$, $\mathbf{P}_{\zeta^2_-}$ (or, perhaps, $\mathbf{P}_{\psi^\omega}$, $\mathbf{P}_{\zeta^1_-}$, $\mathbf{P}_{\zeta^2_+}$). The perfect meshing condition fails, since $\mathbf{P}_{\psi^\omega} \neq \mathbf{P}_{\zeta^1_+} \otimes \mathbf{P}_{\zeta^2_-}$. But the other conditions can be made to hold. By the system representative condition, ω has property $\mathcal{P}_{\psi^\omega}$, 1 has property $\mathcal{P}_{\zeta^1_+}$, and 2 has property $\mathcal{P}_{\zeta^2_-}$. Therefore, by the composition condition, ω has the property $\mathcal{P}_{\zeta^1_+} \otimes \mathcal{P}_{\zeta^2_-}$ corresponding to the projection $\mathbf{P}_{\zeta^1_+} \otimes \mathbf{P}_{\zeta^2_-}$. Now consider $\chi^\omega \;=\; \sin\theta\cdot(\zeta^1_+ \otimes \zeta^2_-) \;+\; \cos\theta\cdot(\zeta^1_- \otimes \zeta^2_+)$, which satisfies $(\psi^\omega, \chi^\omega) = 0$. We have $\mathbf{P}_{\psi\omega} \cdot \mathbf{P}_{\zeta^1_+} \otimes \mathbf{P}_{\zeta^2_-} \cdot \chi^\omega = \cos\theta\sin\theta$ $\cdot\psi^\omega \neq 0$, but $(\mathbf{P}_{\zeta^1_+} \otimes \mathbf{P}_{\zeta^2_-}) \cdot \mathbf{P}_{\psi\omega} \cdot \chi^\omega = 0$. The minimal meshing condition holds, because $\mathbf{P}_{\psi\omega} \cdot (\mathbf{P}_{\zeta^1_+} \otimes \mathbf{P}_{\zeta^2_-}) \cdot \chi^\omega \neq 0$, and $(\mathbf{P}_{\zeta^1_+} \otimes \mathbf{P}_{\zeta^2_-}) \cdot \mathbf{P}_{\psi\omega} \cdot \psi^\omega \neq 0$. But $[\mathbf{P}_{\psi\omega}, \mathbf{P}_{\zeta^1_+} \otimes \mathbf{P}_{\zeta^2_-}] \neq 0$, and so $\mathcal{P}_{\psi\omega}$ and $\mathcal{P}_{\zeta^1_+} \otimes \mathcal{P}\mathbf{P}_{\zeta^2_-}$ are not cogenerable.

Part (b) is proved by reductio ad absurdum. Suppose that two properties $\mathcal{U}^\sigma, \mathcal{V}^\sigma$ ascribed to σ are noncotenable. Consider $\mathcal{U}^\sigma$. If it is composite, then by the composite exclusion condition there must be a composite property $\mathcal{Q}^\sigma \leq \mathcal{U}^\sigma$ which σ has by virtue of the composition condition. Let $\mathcal{Q}^\sigma$ have prime factors $\{\mathcal{R}^{\sigma_i}\}$ ($1 \leq i \leq n$). Then σ_i has $\mathcal{R}^{\sigma_i}$. Since $\mathcal{R}^{\sigma_i}$ is prime, by the prime exclusion condition, either it must be ascribed in accordance with the system representative condition, or there is some composite property $\mathcal{S}^{\sigma_i} < \mathcal{R}^{\sigma_i}$ which σ_i has by virtue of the composition condition. In the latter case, σ has a composite property $\mathcal{S}^\sigma$ with more components than $\mathcal{Q}^\sigma$ such that $\mathcal{S}^\sigma < \mathcal{U}^\sigma$ and $\mathcal{S}^\sigma, \mathcal{V}^\sigma$ are noncotenable. For consider the property $\mathcal{S}^\sigma$ with factors $\{\mathcal{R}^{\sigma_1}, \ldots, \mathcal{S}^{\sigma_i}, \ldots, \mathcal{R}^{\sigma_n}\}$. $\mathbf{S}^{\sigma_i} \cdot \mathbf{P}^{\sigma_i} = \mathbf{S}^{\sigma_i}$, and so $\mathbf{S}^\sigma \cdot \mathbf{Q}^\sigma = \mathbf{S}^\sigma$. Also, $\mathbf{Q}^\sigma = \mathbf{Q}^\sigma$ $\cdot \mathbf{U}^\sigma$. Hence, $\mathbf{S}^\sigma \cdot \mathbf{V}^\sigma = \mathbf{S}^\sigma \cdot \mathbf{Q}^\sigma \cdot \mathbf{U}^\sigma \cdot \mathbf{V}^\sigma = \mathbf{0}^\sigma$ (since $\mathcal{U}^\sigma, \mathcal{V}^\sigma$ are noncotenable). Continuing in this way for a finite number of steps one can arrive at a property $\mathcal{U}'^\sigma < \mathcal{U}^\sigma$ which σ possesses, and whose prime factors $\{\mathcal{U}'^{\sigma'_j}\}$ ($1 \leq j \leq N \geq n$) are all ascribed in accordance with the system representative condition. Now consider the projection $\mathbf{U}^{\star\sigma} = \mathbf{U}^{\star\sigma'_1} \otimes \mathbf{U}^{\star\sigma'_2}$ $\otimes \cdots \otimes \mathbf{U}^{\star\sigma'_N}$, where $\mathbf{U}^{\star\sigma'_j}$ projects onto the individual subspace of σ'_j. We have $\mathbf{U}^{\star\sigma'_j} \cdot \mathbf{U}'^{\sigma'_j} = \mathbf{U}^{\star\sigma'_j}$, since $\mathbf{U}'^{\sigma'_j}$ is ascribed by virtue of the system representative condition. Hence, $\mathbf{U}^{\star\sigma} \cdot \mathbf{U}^\sigma = \mathbf{U}^{\star\sigma}$, and so $\mathbf{U}^\sigma \cdot \mathbf{V}^\sigma = \mathbf{0}^\sigma$ only if $\mathbf{U}^{\star\sigma}$ $\cdot \mathbf{V}^\sigma = \mathbf{0}^\sigma$. Similarly, $\mathbf{U}^\sigma \cdot \mathbf{V}^\sigma = \mathbf{0}^\sigma$ only if $\mathbf{U}^{\star\sigma} \cdot \mathbf{V}^{\star\sigma} = \mathbf{0}^\sigma$, where $\mathbf{V}^{\star\sigma}$ corresponds to $\mathbf{V}^\sigma$ in the same way that $\mathbf{U}^{\star\sigma}$ corresponds to $\mathbf{U}^\sigma$. But it follows from the minimal meshing condition that $\mathbf{U}^{\star\sigma} \cdot \mathbf{V}^{\star\sigma} \neq \mathbf{0}^\sigma$. Hence, $\mathbf{U}^\sigma \cdot \mathbf{V}^\sigma \neq \mathbf{0}^\sigma$, and $\mathcal{U}^\sigma, \mathcal{V}^\sigma$ are cotenable.

That there is no conflict between the subspace decomposition condition and the minimal meshing condition may be shown using the following lemma.

Lemma. Let σ be composed of $\{\sigma_i\}$ and also of $\{\tau_j\}$ ($1 \leq i \leq m$; $1 \leq j \leq n$). For every state ψ^ω of ω,

(a) If there exist accessible subspaces $\mathbf{P}^{\sigma_i} \subseteq \mathsf{H}^{\sigma_i}$, $\mathbf{P}^{\tau_j} \subseteq \mathsf{H}^{\tau_j}$ such that

$$(\mathbf{P}^{\sigma_1} \otimes \mathbf{P}^{\sigma_2} \otimes \cdots \otimes \mathbf{P}^{\sigma_m}) \cdot$$

$$(\mathbf{P}^{\tau_1} \otimes \mathbf{P}^{\tau_2} \otimes \cdots \otimes \mathbf{P}^{\tau_n}) \neq \mathbf{0}, \text{ and}$$

(b) if

$$\psi^\omega = \sum_p b_p^i \cdot (\chi_p^{\sigma_i} \otimes \chi_p^{\bar{\sigma}_i}) \quad (1 \leq i \leq m)$$

$$= \sum_q a_q \cdot (\chi_q^{\sigma} \otimes \chi_q^{\bar{\sigma}})$$

are all biorthogonal expansions of ψ^ω with nonzero coefficients, and privileged sets of subspaces $\{\mathbf{P}_k^{\sigma_i}\}$, $\{\mathbf{P}_k^{\bar{\sigma}_i}\}$, $\{\mathbf{P}_l^{\sigma}\}$, $\{\mathbf{P}_l^{\bar{\sigma}}\}$ [where, e.g., $\mathbf{P}_k^{\sigma_i} \cdot \chi_p^{\sigma_i} = \chi_p^{\sigma_i}$ (for $p \in I_k$) and equals 0 (for $p \notin I_k$)], then, for all i, l

$$\sum_{p \in J_{il}} |b_p^i|^2 \geq \sum_{q \in I_l} |a_q|^2 ,$$

where $p \in J_{il}$ if and only if $(\mathbf{P}_l^{\sigma} \otimes \mathbf{P}_l^{\bar{\sigma}}) \cdot (\mathbf{P}_p^{\sigma_i} \otimes \mathbf{P}_p^{\bar{\sigma}_i}) \neq \mathbf{0}$.

Proof.

(a) Consider the system representative $\mathbf{P}^{\sigma_i}$ of σ_i. This is accessible for σ_i if and only if $\mathbf{P}^{\sigma_i} = \mathbf{P}_k^{\sigma_i}$ (for some k).

255

Now,

$$\left(\sum_k \mathbf{P}_k^{\sigma_i} \right) \cdot \chi_p^{\sigma_i} = \chi_p^{\sigma_i},$$

for all p. Hence,

$$\left[\left(\sum_k \mathbf{P}_k^{\sigma_i} \right) \otimes \mathbf{I}^{\bar{\sigma}_i} \right] \cdot \psi^\omega = \psi^\omega,$$

and so

$$\left[\left(\sum_k \mathbf{P}_k^{\sigma_1} \right) \otimes \left(\sum_l \mathbf{P}_l^{\sigma_2} \right) \otimes \cdots \right] \cdot$$

$$\left[\left(\sum_r \mathbf{P}_r^{\tau_1} \right) \otimes \left(\sum_s \mathbf{P}_s^{\tau_2} \right) \otimes \cdots \right] \cdot \psi^\omega = \psi^\omega.$$

Therefore,

$$\sum_{k,l,\ldots r,s,\ldots} (\mathbf{P}_k^{\sigma_1} \otimes \mathbf{P}_l^{\sigma_2} \otimes \cdots) \cdot (\mathbf{P}_r^{\tau_1} \otimes \mathbf{P}_s^{\tau_2} \otimes \cdots) \cdot \psi^\omega = \psi^\omega.$$

But if, for all accessible subspaces $\mathbf{P}^{\sigma_i}$, $\mathbf{P}^{\tau_j}$,

$$(\mathbf{P}^{\sigma_1} \otimes \mathbf{P}^{\sigma_2} \otimes \cdots \otimes \mathbf{P}^{\sigma_m}) \cdot (\mathbf{P}^{\tau_1} \otimes \mathbf{P}^{\tau_2} \otimes \cdots \otimes \mathbf{P}^{\tau_n}) = \mathbf{0},$$

then, for all $k, l, \ldots, r, s, \ldots$,

$$(\mathbf{P}_k^{\sigma_1} \otimes \mathbf{P}_l^{\sigma_2} \otimes \cdots) \cdot (\mathbf{P}_r^{\tau_1} \otimes \mathbf{P}_s^{\tau_2} \otimes \cdots) \cdot \psi^\omega = 0,$$

and so $\psi^\omega = 0$, which is impossible.

Therefore, part (a) follows by reductio ad absurdum.

(b) We have

$$\sum_{q \in I_l} |a_q|^2 = \left| \left(\psi^\omega, (\mathbf{P}_l^\sigma \otimes \mathbf{P}_l^{\bar{\sigma}}) \cdot \psi^\omega \right) \right|^2$$

$$= \left| \left(\psi^\omega, (\mathbf{P}_l^\sigma \otimes \mathbf{P}_l^{\bar{\sigma}}) \cdot \sum_p b_p^i \cdot (\chi_p^{\sigma_i} \otimes \chi_p^{\bar{\sigma}_i}) \right) \right|^2$$

$$= \left| \left(\psi^\omega, (\mathbf{P}_l^\sigma \otimes \mathbf{P}_l^{\bar{\sigma}}) \cdot \sum_{p \in J_{il}} b_p^i \cdot (\chi_p^{\sigma_i} \otimes \chi_p^{\bar{\sigma}_i}) \right) \right|^2$$

256

$$= \left| \left(\sum_{p' \in J_{il}} b^i_{p'} \cdot (\chi^{\sigma_i}_p \otimes \chi^{\bar{\sigma}_i}_p), (\mathbf{P}^\sigma_l \otimes \mathbf{P}^{\bar{\sigma}}_l) \cdot \right. \right.$$

$$\left. \left. \sum_{p \in J_{il}} b^i_p \cdot (\chi^{\sigma_i}_p \otimes \chi^{\bar{\sigma}_i}_p) \right) \right|^2$$

$$\leq \left| \left(\sum_{p' \in J_{il}} b^i_{p'} \cdot (\chi^{\sigma_i}_p \otimes \chi^{\bar{\sigma}_i}_p), \right. \right.$$

$$\left. \left. \sum_{p \in J_{il}} b^i_p \cdot (\chi^{\sigma_i}_p \otimes \chi^{\bar{\sigma}_i}_p) \right) \right|^2$$

$$\leq \sum_{p \in J_{il}} |b^i_p|^2 \qquad \blacksquare$$

Now there are two ways in which the subspace decomposition condition might possibly come into conflict with the minimal meshing condition. First, it might be that whereas $\{\sigma_i\}, \{\tau_j\}$ compose σ, there are *no* accessible subspaces for the $\{\sigma_i\}, \{\tau_j\}$ such that the minimal meshing condition holds. This possibility is ruled out by part (a) of the lemma. Second, it might be that, for some i, the total probability measure assigned by the subspace decomposition condition to all those accessible subspaces $\{\mathbf{P}^{\sigma_i}_k\}$ for σ_i not ruled out by the minimal meshing condition is still less than the probability assigned to the system representative of σ. In that case, the probability assignments of the subspace decomposition condition could hold only if the minimal meshing condition were to fail. But this possibility is ruled out by part (b) of the lemma, which ensures that the probability assignments of the subspace decomposition condition are always consistent with the minimal meshing condition.

Suppose that a dynamical variable $\mathcal{L}$ with associated operator $\mathbf{L}$ (where $\mathbf{L} = \mathbf{L}^\sigma \otimes \mathbf{I}^\alpha + \mathbf{I}^\sigma \otimes \mathbf{L}^\alpha$) is conserved, in the sense that $[\mathbf{L}, \mathbf{U}_{t_0 t_1}] = \mathbf{0}$, where $\mathbf{U}_{t_0 t_1}$ is the evolution operator corresponding to Schrödinger evolution of a vector over the duration (t_0, t_1) of an interaction which is M-suitable for $\mathscr{A}$.

Theorem. $[\mathbf{L}^\sigma, \mathbf{A}^\sigma] = 0$.

Proof. Since $\mathbf{U}_{t_0 t_1}$ is the evolution operator corresponding to Schrödinger evolution of a vector over the duration (t_0, t_1) of an interaction which is M-suitable for $\mathscr{A}$, $\mathbf{U}_{t_0 t_1}$ has the following effect on certain vectors in $\mathsf{H}^\sigma \otimes \mathsf{H}^\alpha$:

$$\mathbf{U}_{t_0 t_1} (\chi_{ij}^\sigma \otimes \chi_0^\alpha) = (\xi_{ij}^\sigma \otimes \chi_{ij}^\alpha),$$

where

$$(\xi_{ij}^\sigma, \xi_{i'j'}^\sigma) = (\chi_{ij}^\alpha, \chi_{i'j'}^\alpha) = \delta_{ii'} \cdot \delta_{jj'}.$$

Now

$$\left(\chi_{i'j'}^\sigma \otimes \chi_0^\alpha , \mathbf{L}(\chi_{ij}^\sigma \otimes \chi_0^\alpha) \right)$$

$$= \left(\mathbf{U}_{t_0 t_1}(\chi_{i'j'}^\sigma \otimes \chi_0^\alpha) , \mathbf{U}_{t_0 t_1}\mathbf{L}(\chi_{ij}^\sigma \otimes \chi_0^\alpha) \right)$$

$$= \left(\mathbf{U}_{t_0 t_1}(\chi_{i'j'}^\sigma \otimes \chi_0^\alpha) , \mathbf{L}\mathbf{U}_{t_0 t_1}(\chi_{ij}^\sigma \otimes \chi_0^\alpha) \right)$$

$$= \left(\xi_{i'j'}^\sigma \otimes \chi_{i'j'}^\alpha , \mathbf{L}(\xi_{ij}^\sigma \otimes \chi_{ij}^\alpha) \right)$$

$$= \left(\xi_{i'j'}^\sigma \otimes \chi_{i'j'}^\alpha , (\mathbf{L}^\sigma \otimes \mathbf{I}^\alpha + \mathbf{I}^\sigma \otimes \mathbf{L}^\alpha) (\xi_{ij}^\sigma \otimes \chi_{ij}^\alpha) \right)$$

$$= (\xi_{i'j'}^\sigma, \mathbf{L}^\sigma\xi_{ij}^\sigma) (\chi_{i'j'}^\alpha, \chi_{ij}^\alpha) + (\xi_{i'j'}^\sigma, \xi_{ij}^\sigma) (\chi_{i'j'}^\alpha, \mathbf{L}^\alpha\chi_{ij}^\alpha)$$

$$= 0, \text{ unless } i = i'.$$

It follows that $(\chi_{i'j'}^\sigma, \mathbf{L}^\sigma\chi_{ij}^\sigma) = 0$, unless $i = i'$. Now, write $\mathbf{A}^\sigma = \Sigma_i a_i \mathbf{P}_i^\sigma$.
Then we have

$$\left(\chi_{i'j'}^\sigma , \mathbf{L}^\sigma\mathbf{P}_i^\sigma\chi_{i''j}^\sigma \right)$$

258

$$= \delta_{ii''} \left(\chi^{\sigma}_{i'j'} , \mathbf{L}^{\sigma}\chi^{\sigma}_{i''j} \right)$$

$$= \delta_{ii''} \left(\chi^{\sigma}_{i'j'} , \mathbf{L}^{\sigma}\chi^{\sigma}_{ij} \right)$$

$$= \delta_{ii'} \delta_{ii''} \left(\chi^{\sigma}_{i'j'} , \mathbf{L}^{\sigma}\chi^{\sigma}_{ij} \right)$$

$$= \delta_{ii'} \delta_{i'i''} \left(\chi^{\sigma}_{i'j'} , \mathbf{L}^{\sigma}\chi^{\sigma}_{ij} \right)$$

$$= \delta_{ii'} \left(\chi^{\sigma}_{i'j'} , \mathbf{L}^{\sigma}\chi^{\sigma}_{i''j} \right)$$

$$= \left(\chi^{\sigma}_{i'j'} , \mathbf{P}^{\sigma}_{i}\mathbf{L}^{\sigma}\chi^{\sigma}_{i''j} \right)$$

Hence, $[\mathbf{L}^{\sigma}, \mathbf{P}^{\sigma}_{i}] = \mathbf{0}$, and so $[\mathbf{L}^{\sigma}, \mathbf{A}^{\sigma}] = \mathbf{0}$. ∎

Selected bibliography

Araki, H. and Yanase, M. M. (1960) "Measurement of Quantum Mechanical Operators," *Physical Review* **120**, 622–6.

Aspect, A., Grangier, P., and Roger, G. (1982a) "Experimental Realization of Einstein-Podolsky-Rosen-Bohm *Gendankenexperiment:* a New Violation of Bell's Inequalities,"*Physical Review Letters* **49**, 91–4.

Aspect, A., Dalibard, J., and Roger, G. (1982b) "Experimental Tests of Bell's Inequalities Using Time-Varying Analyzers," *Physical Review Letters* **49**, 1804–7.

Ballentine, L. (1970) "The Statistical Interpretation of Quantum Mechanics," *Reviews of Modern Physics* **42**, 358–81.

Bell, J. S. (1964) "On the Einstein Podolsky Rosen Paradox," *Physics* **1**, 195–200.

   (1987) *Speakable and Unspeakable in Quantum Theory.* Cambridge: Cambridge University Press.

Bohm, D. (1951) *Quantum Theory.* Englewood Cliffs, NJ: Prentice Hall.

   (1952a) "A Suggested Interpretation of the Quantum Theory in Terms of 'Hidden Variables': Part I," *Physical Review* **85**, 166–79.

   (1952b) "A Suggested Interpretation of the Quantum Theory in Terms of 'Hidden Variables': Part II," *Physical Review* **85**, 180–93.

Bohr, N. (1963) *Essays 1959–62 on Atomic Physics and Human Knowledge.* New York: Interscience.

Bub, J. (1974) *The Interpretation of Quantum Mechanics.* Dordrecht: Reidel.

Carnap, R. (1939) *The Foundations of Logic and Mathematics,* 4:3 of the *International Encyclopedia of Unified Science,* O. Neurath, R. Carnap, and C. Morris, eds. Chicago: University of Chicago Press, 1955.

Cartwright, N. (1988) "Quantum Causes: the Lesson of the Bell Inequalities," paper read at the Joint US-USSR Colloquium on the Foundations of Quantum Mechanics, Easton, Maryland.

Davies, P. C. W. and Brown, J. R., eds. (1986) *The Ghost in the Atom.* Cambridge: Cambridge University Press.

D'Espagnat, B. (1979) "The Quantum Theory and Reality," *Scientific American* **241**, 158–81.

DeWitt, B. and Graham, N., eds. (1973) *The Many Worlds Interpretation of Quantum Mechanics*. Princeton: Princeton University Press.

Dummett, M. (1978) "Is Logic Empirical?," in *Contemporary British Philosophy, 4th Series,* H. D. Lewis, ed., reprinted in *Truth and Other Enigmas*. London: Duckworth Press, 1978.

Earman, J. and Norton, J. (1987) "What Price Spacetime Substantivalism?," *British Journal for the Philosophy of Science* **38**, 515–25.

Einstein, A. (1949) "Reply to My Critics," in Schilpp, ed. (1949).

Einstein, A., Podolsky, B., and Rosen, N. (1935) "Can Quantum-Mechanical Description of Physical Reality Be Considered Complete?," *Physical Review* **47**, 777–80.

Everett, H. (1957) " 'Relative State' Formulation of Quantum Mechanics," reprinted in DeWitt and Graham, eds. (1973).

    (1973) "Theory of the Universal Wave Function," in DeWitt and Graham, eds. (1973).

Fine, A. (1970) "Insolubility of the Quantum Measurement Problem," *Physical Review* D2, 2783–7.

Finkelstein, D. (1962) "The Logic of Quantum Physics," *Transactions of the New York Academy of Science* **25** (1965), 621–37.

Fleming, G. (1985) (unpublished) "Towards a Lorentz Invariant Quantum Theory of Measurement." Invited lectures presented at the Minicourse and workshop on Fundamental Physics, Universidad de Puerto Rico.

Friedman, M. (1983) *Foundations of Space-Time Theories*. Princeton: Princeton University Press.

Friedman, M. and Putnam, H. (1978) "Quantum Logic, Conditional Probability and Interference," *Dialectica* **32**, 305–15.

Geroch, R. (1984) "The Everett Interpretation," *Nous* **18**, 617–33.

Gleason, A. M. (1957) "Measures on the Closed Subspaces of a Hilbert Space," *Journal of Mathematics and Mechanics* **17**, 59–81.

Greenberger, D. (1983) "The Neutron Interferometer as a Device for Illustrating the Behavior of Quantum Systems," *Review of Modern Physics* **55**, 875–905.

Healey, R. A. (1977) (unpublished) "Quantum Mechanics, Realism and Quantum Logic."

    (1979) "Quantum Realism: Naivete Is No Excuse," *Synthese* **42**, 121–44.

    (1981) "Comments on Kochen's Specification of Measurement Interactions," in *PSA 1978, Volume 2,* Asquith, P. D. and Hacking, I., eds. East Lansing, MI: Philosophy of Science Association.

    (1984a) "How Many Worlds?," *Nous* **18**, 591–616.

    (1984b) "On Explaining Experiences of a Quantum World," in *PSA*

1984, Volume 1, Asquith, P. D. and Kitcher, P., eds. East Lansing, MI: Philosophy of Science Association, 56–69.

Howard, D. (1985) "Einstein on Locality and Separability," *Studies in History and Philosophy of Science* **16**, 171–201.

Jarrett, J. (1984) "On the Physical Significance of the Locality Conditions in the Bell Arguments," *Nous* **18**, 569–89.

Kochen, S. (1978) (unpublished) "A New Interpretation of Quantum Mechanics."

 (1985) "A New Interpretation of Quantum Mechanics," *Symposium on the Foundations of Modern Physics,* Lahti, P. and Mittelstaedt, P., eds. Teaneck, NJ: World Scientific Publishing Co., 151–70.

Kochen, S. and Specker, E. P. (1967) "The Problem of Hidden Variables in Quantum Mechanics," *Journal of Mathematics and Mechanics* **17**, 59–81.

Leggett, A. J. (1980) "Macroscopic Quantum Systems and the Quantum Theory of Measurement," *Supplement of The Progress of Theoretical Physics* **69**, 80–100.

Lewis, D. (1986) "Causation," in *Philosophical Papers, Volume 2,* Lewis, D., ed. Oxford: Oxford University Press.

Mackey, G. W. (1963) *Mathematical Foundations of Quantum Mechanics.* New York: Benjamin.

Margenau, H. (1963) "Measurements and Quantum States," *Philosophy of Science* **30**, 1–16, 135–157.

Norman, S. (1981) (unpublished) "Subsystem States in Quantum Theory and Their Relation to the Measurement Problem," Stanford University Ph.D. thesis.

Putnam, H. (1968) "Is Logic Empirical?," in *Boston Studies in the Philosophy of Science, Volume 5,* Cohen, R. and Wartofsky, M., eds., reprinted as "The Logic of Quantum Mechanics," in *Philosophical Papers, Vol. 1,* Putnam, H., ed. Cambridge: Cambridge University Press, 1975.

Putnam, H. (1981) "Quantum Mechanics and the Observer," *Erkenntnis* **16**, 193–219; reprinted in *Philosophical Papers, Vol. 3,* Putnam, H., ed. Cambridge: Cambridge University Press, 1983, 248–70.

Redhead, M. (1987) *Incompleteness, Nonlocality, and Realism: a Prolegomenon to the Philosophy of Quantum Mechanics.* Oxford: Clarendon Press.

Reichenbach, H. (1956) *The Direction of Time,* Reichenbach, M., ed. Berkeley: University of California Press.

Salmon, W. C. (1984) *Scientific Explanation and the Causal Structure of the World.* Princeton, NJ: Princeton University Press.

Schilpp, P. A., ed. (1949) *Albert Einstein: Philosopher-Scientist.* La Salle, IL: Open Court.

263

Schrödinger, E. (1935a) "Die Gegenwärtiger Situation in der Quanten-mechanik," *Die Naturwissenschaften* **23**, 807–12, 823–8, 844–9.

(1935b) "Discussion of Probability Relations Between Separated Systems," *Proceedings of the Cambridge Philosophical Society* **31**, 555–63.

Shimony, A. (1974) "Approximate Measurement in Quantum Mechanics, II," *Physical Review* **D9**, 2321–3.

(1986) "Events and Processes in the Quantum World," in *Quantum Concepts in Space and Time*, Penrose, R. and Isham, C. J., eds. Oxford: Clarendon Press, 182–203.

Stairs, A. (1983) "Quantum Logic, Realism, and Value Definiteness," *Philosophy of Science* **50**, 578–602.

(1988) "Jarrett's Locality Condition and Causal Paradox," in *PSA 1988, Volume 1*, Fine, A. and Leplin, J., eds. East Lansing, MI: Philosophy of Science Association, 318–25.

Stein, H. (1972) "On the Conceptual Structure of Quantum Mechanics," in *Paradigms and Paradoxes*, Colodny, R. G., ed. Pittsburgh, PA: University of Pittsburgh Press, 367–438.

(1984) "The Everett Interpretation of Quantum Mechanics: Many Worlds or None?" *Nous* **18**, 635–52.

Stein, H. and Shimony, A. (1972) "Limitations on Measurement," in *Proceedings of the International School of Theoretical Physics "Enrico Fermi,"* D'Espagnat, B., ed. New York: Academic Press, 56–75.

Van Fraassen, B. C. (1980) *The Scientific Image*. Oxford: Clarendon Press.

(1981) "A Modal Interpretation of Quantum Mechanics," in *Current Issues in Quantum Logic*, Beltrametti, E. G. and Van Fraassen, B. C., eds. New York: Plenum, 229–58.

Vigier, J. P. (1982) "Non-locality, Causality and Aether in Quantum Mechanics," *Astronomische Nachrichten* **303**, 55–80.

von Neumann, J. (1932) *Mathematische Grundlagen der Quantenmechanik*. Berlin: Springer; reprinted as *Mathematical Foundations of Quantum Mechanics*, trans. by R. T. Beyer. Princeton NJ: Princeton University Press, 1955.

Wheeler, J. A. (1978) "The 'Past' and the 'Delayed-Choice' Double Slit Experiment," in *Mathematical Foundations of Quantum Theory*, Marlow, A. R., ed. New York: Academic, 9–48.

Wigner, E. P. (1952) "Die Messung Quantenmechanischer Operatforen," *Zeitschrift für Physik* **133**, 101.

(1963) "The Problem of Measurement," *American Journal of Physics* **31**, 6–15.

Yanase, M. M. (1961) "Optimal Measuring Apparatus," *Physical Review* **123**, 666–8.

Index

Note: entries for terms and principles defined in the text appear in **bold face type**, as do those pages of the text where they are defined.

accessible dynamical state, 29, 41, 45–6, 104
accessible subspace, **77–8**, 80, 92, 96–7, 110, 113, 231, 237, 256–7
action at a distance, 8*fn*, 50, 59 (*see also* NIAD)
Albert D., xiii, 244
anticorrelation, 49–50, 53, 55, 120, 153–7, 228
apparatus (*see* system)
approximate SMI, **102**, 245–6
Araki, H., 101
Aspect, A., xi, xii, 52, 55, 116, 135, 140–1, 144–5
atomic system (*see* system)

backtracking argument, **198**, 203–5
Ballentine, L., 3*fn*,
Bell, J. S., xii, 16–17, 22–3, 47, 49, 116, 166, 182, 216*fn*, 217
Bell inequalities, 3*fn*, **47–8**, 50, 54, 58–60, 128–9, 134, 150, 242
biorthogonal decomposition lemma, **77**, 214, 218, 221–2, 231, 237
Bohm, D., 3*fn*, 22, 24*fn*, 48, 50–2, 116–19, 128, 131, 195–6, 224, 226
Bohr, N., 1–3, 10, 13*fn*, 16*fn*, 37, 217
Boolean algebra, 64
Boolean lattice, 73
Boolean σ-algebra, 221–3, 231
Borel set, 7, 66, 74

Born rules, xi, xii, 10–13, 20, 27–31, 35–41, 44–8, 52, 58, 64, 90, 105–8, 123, 129, 150, 152–3, 156, 167–9, 172, 186, 193, 196–7, 212, 218, 233
  derivation of, 45, 67, 84–5, 105–6, 109–14, 215, 234, 243–4
  interpretation of, 5, 7–11, 14, 21, 32–3, 47, 84, 180–1, 218, 224
Brown, J. R., 2*fn*
Bub, J., xiii, 3*fn*, 21
Butera, L., xii

Carnap, R., 4*fn*
Cartwright, N., xiii, 178
causality, 37, 145, 163, 177–8, 225, 228
  connection, 50–1, 55, 57, 59–61, 137, 145–6, 176–7
  explanation, xii, 57, 115, 137, 152, 154–64, 168–9, 172–3, 176, 178
  future, 139, **142**–3, 151–2, 156–9, 177, 197, 240
  influence, 23, 132
  interaction, 53
  paradox, 23, 60, 146–8
  past, **142**–3, 147–8, 156, 197, 240
  process, 59–62, 160–5, 168–70, 176–8
  relation, 57, 60–1, 159, 176–7
  signal, 23, 146
causation, 37, 48, 60, 159, 173, 176–8
cause, 51, 59–60, 177–9
  common, 59–61, 178
characterization problem, **188**–93

classic, 14, 35–6, 42*fn*, 84, 139
  conception of measurement, 39–41
  description, 15, 37, 47
  mechanics, 10, 15, 22–4, 28, 31,
    164–5, 183
  state (*see* state)
  system (*see* system)
cogenerable, 73–4, 253
complement, 64, 241
completeness, 17, 19, 22, 28, 36, 183–4
  Jarrett, 58–9
composes, 64
composite, 22, **65**–71
composite exclusion condition, 69,
  253–4
composition condition, 69, 121,
  253–4
Copenhagen interpretation, 1–3, 6, 9–
  10, 13, 16, 17–19, 22, 24*fn*, 48,
  52*fn*, 138, 172, 180, 184–205
  strong version, 13–16, 168, 191–7
  weak version, 10–13, 28, 48, 148–
    51, 169, 186–91, 198–9
cotenable, 73–4, 253–5
covariance
  general, 139–40, 144
  Lorentz, 139, 238
  relativistic, 139

Dalibard, J., xi,
Davies P.C.W., 2*fn*,
de Broglie, L., 22,
definite result (*see* determinate
  outcome)
degenerate, 77, 98–102, 127–8
  apparatus dynamical state, 98–9,
    102
delayed-choice experiments, 36–7, 202
D'Espagnat, B., 168
determinate, 73
  outcome, 39, 88, 98–104, 127, 169,
    187, 192, 203–5, 210–14, 222–3,
    246
DeWitt, B., 3*fn*, 206*fn*, 216*fn*
dichotomy, 74
Dirac, P.A.M., 238–41
disjoint, 64
dispositions, 45, 47, 55, 130, 155–8,
  177–8, 230
  conditional, 54–5, 131–4
  probabilistic, xi, 31, 54–**5**, 104–5,
    131–4, 156–8, 185–6, 234

  unconditional, 54–5, 57, 133–4
Dummett, M., 21*fn*
dynamics, 14, 22, 30, 42*fn*, 45, 79–83,
  104, 183–4, 221, 231*fn*, 235–8

Einstein A., 3*fn*, 23, 50–2, 149, 181*fn*
  (*see also* EPR)
Einstein locality, 50–4, 58, 150, 219
Einstein separability, 51–2
elementary sentence, 27–8, 66
ensemble, 13–14, 22, 27, 38, 46–7,
  105, 107, 114, 185, 192, 196–7,
  201–3, 234
environment, 95–6, 99–100, 108, 111–
  12, 127, 129, 188, 201–3, 220–2,
  230, 233, 248
EPR, xi, xii, 6, 48–52, 116–20, 131–4,
  137, 143, 145, 151–2, 162–78,
  182–3, 195–6, 216–19, 224–6,
  229–30, 233
Everett, H. III, 3, 19, 205–18, 232–3
evolution, 30, 38, 42*fn*, 79–82, 86–7,
  91, 93, 96–7, 101, 103, 109,
  112, 121–30, 183–4, 188, 195,
  235–40, 247
experimental arrangement, 16, 34, 37,
  185, 192–4, 244
explanation
  of definite outcomes according to
    Everett interpretation, 210
  of EPR-type correlations, 52, 54–7,
    131–4, 137, 143, 152–60, 164–
    76, 178, 218–19, 229, 242
  incompleteness of Copenhagen
    interpretation, 17–19, 195–7
  of legitimacy of applying quantum
    mechanics to interacting sys-
    tems, 221
  of role of Stern-Gerlach device, 202
  of stability of dynamical properties,
    223
  of success of projection postulate,
    206, 208, 233–4

factorizability, 65, 69
Fine, A., 12*fn*
Finkelstein, D., 3*fn*,
Fleming, G., 150*fn*, 197, 239*fn*
free evolution, 80, 127, 239–40
Freidman, M., 3*fn*, 60*fn*, 146*fn*

Geroch, R., xiii, 3*fn*
Gleason, A. M., 8*fn*, 182

good observation, **206**–10, 213–16
Graham, N., 3*fn*, 206*fn*
Grangier, P., xi
Greenberger, D., 200*fn*

Hampton, J., xiv
happening, 61–2, 175–**6**
Healey, R. A., 3*fn*, 8, 22*fn*, 182, 206*fn*, 210*fn*, 211*fn*, 212*fn*, 216
Heisenberg, W., 11, 16
Hellman, G., xiii, 198*fn*
hidden variables, 2–4, 21, 24–5, 47, 181–2, 184
   nonlocal, 3, 23, 132
hierarchy, 29–30, 63–6, 68, 76, 241
Hilbert space, 5, 53, 64–5, 68–73, 77–8, 93, 98, 111, 116–17, 237
holism, xii, 29, 48, 61, 116, 143, 173, 191, 229, 242
holistic, 62, 174–5
holistic$_1$, **175**–6
holistic$_2$, **175**–6, 179
holistic dynamical state, **174**–5, 181
holistic explanation, 137, **174**–5
holistic process, 59, **143**–4, 164–6, **174**–5
holistic stage, **174**–5
Howard, D., 50*fn*
hyperplane, 138–45, 150, 155–8, 162–3, 178, 197, 239–40
hypersurface, 135–6, 138, 141, 144, 163, 178, 239

indeterminacy relations (*see* Heisenberg)
interaction, xi, 12, 19–20, 30–45, 50–4, 77–84, 97, 99–104, 109, 112, 116, 127–9, 135–6, 138–9, 143, 160, 162–3, 170, 172, 181, 184, 186–91, 193, 201–7, 209–10, 212–13, 218–30, 234–8, 242, 248 (*see also* measurement, preparation)
interactive interpretation, **33**
   open questions for, 235–51
interpretation, xi–xiv, 1–7, 24, 26–35, 37–8, 41, 44–5, 48, 52, 58, 60, 84, 90, 101, 104, 106, 115, 134–5, 137–40, 167–8, 171–4, 176, 179 (*see also* Copenhagen interpretation; naive realism; quantum logic; Everett, H.)

intersubjective reproducibility of measurements (IRM), **208**–10, 215
invariance
   Lorentz, 140, 144–5, 219, 238–9
   relativistic, 24, 139, 145, 149, 151, 169–70, 183, 190–2, 197, 230
irreducible property, **29**, 52–4, 61–2, 65–6, 68, 72, 88, 116, 121, 131, 143, 164–7, 170–6, 191, 218–9, 229

Janis, A., xiii, 217*fn*
Jarrett, J., 23*fn*, 58, 148

Klein-Gordon equation, 238–9
knowledge, 9, 38, 43, 104–5, 164–73, 185, 223–9, 234, 249
Kochen, S., 8*fn*, 77, 120, 128, 216–34
Kripke, S., 21*fn*

Leggett, A.J., 42*fn*
Lewis, D., 157
locality, 23, 47–8, 52, 58, 62, 138–9, 144, 160, 166, 182–3, 230, 242
   Bell, **50**–1, 58
   Einstein, **50**–1, 150, 219
   strong, 58
local spacetime theories, 139–40, 144
Lorentz, H.A., 139–40, 149, 238–9 (*see also* covariance, invariance)
   boost, 144–5

macroscopic, 15, 17, 32, 195, 248 (*see also* property, system)
many-worlds interpretation, 2–4, 19–21, 211–12
Margenau, H., 199–202
mark transmission, 160–4
Martin, D.A., xiv
measurement, xi, 6, 8–11, 15, 19–21, 27, 30–3, 36–7, 39–45, 49–53, 58, 74, 84, 87–90, 95–6, 102–3, 106, 114–16, 135, 150–1, 164–5, 171, 180–1, 184–6, 188–9, 192–3, 195–204, 206–10, 212, 214–15, 219, 226–7, 229, 232, 243–4
   approximate, 101–3, 244–8
   classical conception (*see* classic)
   device, 37, 100, 156, 170
   ideal, 44, 101, 111, 114, 193
   interaction, xii, 9–11, 14–16, 20,

267

measurement (*cont.*)
  39, 41, 43–4, 52, 54–5, 59, 66,
  82–3, 84–96, 98, 100, 116, 170,
  172, 181, 185–8, 191, 194, 202–
  6, 211–14, 218, 222–3, 225,
  227–9, 231–2, 242–3, 250
  outcome, 15, 20, 22, 44, 58, 83–4,
  90–1, 93, 99, 106–8, 181, 185,
  196, 199, 206, 210, 212, 215,
  223, 233–4, 248, 250–1
  problem, xi, **11**–13, 15, 41–2, 44,
  99, 172–3, 187, 192, 198, 216,
  242
  preparatory, 188
  result (*see* measurement outcome)
  type interaction, 42–4, 53, 91, 106,
  184, 201, 223, 228, 201–2, 223,
  245
mechanism, 17, 153
meshing condition
  minimal, 72–3, **79**, 253–5, 257
  perfect, **72**–3, 76, 78–9
minimal set value, **74**–5
minimally disturbing SMI, **91**–2,
  121, 125, 127
Minkowski spacetime, 141
M-ready interaction, **86**–7, 93
M-suitable interaction, **86**–7, 93, 101–
  3, 224*fn*, 258
M-type interaction, 31–6, 39, 45, **84**–5,
  86, 102–8, 172, 201, 227–8, 234,
  243, 245–8
M-type system, 35–6, 86–**6**, **87**–8,
  93–4, 170

naive realism, 2–4, 8–9, 16, 21, 28, 49,
  51, 57, 131, 135, 138, 148, 150–
  1, 169, 180–4
negation, **72**
neutron interferometer, 200
Newton, I., 138, 140, 171, 250
NIAD, **50**–1, 53–4, 57, 59–61, 141,
  145–6, 148, 151, 159
nonlocality (*see* hidden variables,
  locality)
nonseparability, 29, 66, 137, 141–4,
  162, 168*fn*, 182–3, 230, 242
Norman, S., 216, 224
null subsystem, **83**

object system, *see* system
observation, *see* measurement

operator, 5, 27, 64, 87–8, 101, 115,
  117, 121, 206–7, 215, 221, 231,
  238, 246, 257
  Hamiltonian, 30, 77–81, 85–6, 91,
  93, 108–9, 112, 206–7, 215,
  220–1
  projection, 27, 65, 72, 92, 110, 115,
  221, 253
  unitary, 80–2, 97, 101
outcome independence, 148

parameter independence, 148, 177
Pauli, W., 1–3
Peierls, R., 1–4
Podolsky, B., *see* EPR
pointer position, 13, 39, 187, 208–10,
  213–15, 218, 222–3
polar decomposition theorem, 218,
  230–1
precise value, **8**–11, 16, 21, 28, 40–2,
  73–7, **74** , 88, 180, 182, 184,
  221–2
preparation, 31, 34, 37, 104–15, 199–
  205, 215, 227–8, 233–4, 243
  interaction, xii, 34–5, 37–8, 43,
  105–10, 200, 215, 234, 243
  problem, **187**–8, 192
prime, **65**, 68–71, 87, 92–4
prime exclusion condition, **70**,
  253–4
prime factor, **65**, 67, 69, 254
privileged set, **78**, 80–1, 89, 221, 255
probabilistic law of coexistence, **46**
probability, 17, 30–1, 38–9, 45–7, 67,
  82–3, 98, 105, 147–8, 157, 177–
  9, 181, 185, 208, 220–1, 224–8,
  232–4, 250–1 (*see also* Born
  rules)
process, 17–18, 22, 42–3, 61, 116, 131,
  138–41, 147–9, 151–2, 161, 173,
  188, 190–2, 200, 206, 211–13,
  216, 235, 239, 250–1
  causal (*see* causality process)
  continuous, 154–6, 158, 162, 178,
  236
  holistic, 59, **143**–4, 164–6, **174**–5
  indeterministic, 42, 45, 133, 186,
  236
  nonseparable, **142**–5, 153, 155,
  157–60, 162–6, 170, 176, 182–3,
  230, 233
  pseudo-, 141, 159–60

projection postulate, 5, **10**–11, 42–4, **89–92**, 106, 111, 114, 188–90, 196–7, 199–200, 202, 205–10, 214–16, 223, 233
property, 13–14, 20, 28, 49, 53, 92, 96, 99, 115, 131, 134–5, 144, 149–53, 163, 171, 174, 197, 229
  composite, 65, 67–71, 87–8, 92–4, 121, 254
  correlational, 53–4, 121–3, 126–31, 133, 165–6, 229
  dynamical, 64–6, 68, 74–6, 79, 186, 218, 222–5, 227, 231–2
  irreducible, 29, 52, 54, 61–2, 65–8, 72, 79, 131, 143, 151, 164–7, 170–6, 219, 229
  macroscopic, 15
  prime, 65–71, **87–92**
  reducible, 29, 54, 65–6, 79, 96, 132
  recording, 11–12, 16, 40, 42, 44, 53–4, 86–8, 91, 94, 99
  spatial, 138, 141
property compatibility, 74
property inclusion condition, 67–8
property intersection, 74
proviso, 70–1, 73, 88, 92, 94
P-type interaction, 31–6, 45, 108, 111–12, 234, 243–4
Putnam, H., xiii, 3*fn*, 21, 22*fn*

quantum field theory, 238, 241–3
quantum logic, 2–4, 21–22

rank, 64, 66, 68, 71, 76
realism, 6, 7*fn*, 149, 211, 230 (*see also* naive realism)
recording quantity, 87, 244, 246–7
Redhead, M., 8*fn*, 163, 182
Reichenbach, H., 160, 178
relativistic quantum theory, 238–42
relativity, principle of, 8*fn*, 23–24, 149, 190
Rescher, N., xii
Roger, G., xi,
Rosen, N., *see* EPR

Salmon, W.C., 157, 160–2, 178
Schilpp, P.A., 181*fn*
Schmidt, 77
Schrödinger, E., 50, 79, 187
Schrödinger equation, 30, 42*fn*, 79, 86,

90, 101, 188, 190, 203–4, 206–8, 237, 239, 258
Schrödinger's cat, 6, 12–13, 15, 20
separability (*see* Einstein, nonseparability)
set value, 74
Shimony, A., xiii, 12*fn*, 101, 148
signal, 58
  first, 146
simple M-type interaction (SMI), 85–7, 88–9, 91–5, 101–3, 109, 111–14, 244
simple preparation procedure (SPP), 107, 114, 244
  type 1, 108
  type 2, 111–2, 201
spacelike property, 49–52, 55–62, 135, 137–42, 144–5, 147–51, 155–8, 169, 177–8, 191, 197, 239
spatiotemporal continuity, 154–5, 158, 162, 178
Specker, E. P., 8*fn*, 182
stability condition, 82, 110, 127, 215–16, 222–3, 234, 236, 239–40
Stairs, A., xiii, 22*fn*, 148
Stapp, H. P., 17
state, 26–47
  classical, 14–15, 31, 41–2, 47, 183–4, 248
  definite, 13–14, 28, 42, 103–4, 192, 198–9, 203–5, 210, 212–14, 219, 222–3, 245, 248
  dynamical, xi–xii, 15, 21, 27–31, 34–47, 50–2, 59, 61, 63–83, 85, 95, 103–5, 111, 114–15, 129, 138, 143, 148, 163, 169, 174–5, 180–6, 190–2, 195–6, 205, 218–22, 224, 228–9, 230–40, 242, 245, 247–50
  ghost, 225, 228
  outcome, 32–3, **39,** 41–2
  quantum, xi–xii, 6–7, 10–12, 14–15, 20, 27–8, 30–8, 41–5, 53, 64, 68, 79, 83–5, 89, 104–15, 148–9, 168–9, 180–2, 185–99, 204, 207–8, 210, 212–13, 218, 225–6, 233–4, 238–40, 242
  real, 50–1, 52*fn*, 150
  relative, 19, 20, 212–13
  set, 29
  singlet, 48, 52, 116, 129, 152, 167, 186, 190, 196–7

state (*cont.*)
    vector, 27, 29, 86, 90, 114–15, 123,
        207, 213, 231, 233
Stein, H., xiii, 3*fn*, 65, 87, 101, 210*fn*
Stern-Gerlach device, 34–8, 43, 114,
    198–204, 244
**subjective reproducibility of mea-
    surements (SRM)**, 207, 214–15
subspace decomposition condition,
    77–**8** , 79–81, 135, 214, 218,
    220–1, 234, 236–7, 240–1, 249,
    253–5, 257
system, xi, 15
    apparatus, xi, 5, 11–18, 20, 32,
        36–7, **39**–42, 44, 47, 53, 85, 89–
        91, 93, 95, 98–9, 102–4, 106,
        108–9, 111–12, 183, 187, 192–5,
        201, 203–4, **206**–11, 213, 218,
        222–3, 225–6
    atomic, **63**–4, 68–9, 73, 78, 87,
        116, 184*fn*, 241, 249, 253
    classical, 9–10, 19, 31, 39, 185, 188,
        206, 214, 248
    compound, xi, 19–20, 22, 39, 45–7,
        51, 53–4, 62, 71–2, 116–17,
        130–2, 170–2, 174–6, 181–3,
        191, 205–6, 208, 213, 219, 229,
        233
    coupled, xii, 116–36
    macroscopic, 12–14, 39, 42, 76, 189
    object, 11, 16, 20, 22, **39**–41, 44,
        85, 89–91, 95, 98, 100, 188–9,
        193–4, 203, 206–7, 225–6
    probe, 39–40
    representative, 29–30, **68**, 70–3,
        76–83, 86–9. 91–102, 110–17,

        120–8, 135, 220, 222, 231–2,
        237–40, 245, 253, 256–7
    representative condition, **68–9**,
        70, 73, 88, 92, 94, 96, 110, 113,
        230*fn*, 232, 236, 253–4
    universal, **63**, 67, 77, 79, 87, 89,
        91, 95–6, 98–100, 109, 113, 135,
        220, 241, 249–50

timelike property, 52, 54–6, 59, 149,
    177, 149, 177
Torretti, R., xiii

uncorrelated system, **95**–6
    strongly, **96**–8
universe, 19–20, 41, 63, 78, 187–8,
    192, 206, 212, 249

van Fraassen, B.C., 7*fn*, 170*fn*
**verifiability of measurement results
    (VMR)**, 90, **209**
Vigier, J.P., 3*fn*, 22–3
von Neumann, J., 2, 5, 77, 89

wave
    function, 27, 239
    packet, 42–4, 111, 186, 247
weakening condition, **67**–70, 96,
    123, 253
weaker than property, **67**, 69–70, 88,
    92, 94
Wigner, E., 2, 12*fn*, 34, 99, 101, 199–
    203
witnessed state, **230**–3

Yanase, M.M., 101

270

Printed in the United States
By Bookmasters